Darwin and Theories of Aesthetics and Cultural History

Edited by
Barbara Larson and Sabine Flach

Routledge
Taylor & Francis Group
LONDON AND NEW YORK

First Published 2013 by Ashgate Publisher

Published 2016 by Routledge
2 Park Square, Milton Park, Abingdon, Oxfordshire OX14 4RN
711 Third Avenue, New York, NY 10017, USA

First issued in paperback 2016

Routledge is an imprint of the Taylor & Francis Group, an informa business

British Library Cataloguing in Publication Data
Darwin and theories of aesthetics and cultural history.
1. Darwin, Charles, 1809–1882—Influence. 2. Aesthetics, Modern. 3. Art and science.
I. Larson, Barbara (Barbara Jean) II. Flach, Sabine.
701'.05—dc23

Library of Congress Cataloging-in-Publication Data
Darwin and theories of aesthetics and cultural history / edited by Barbara Larson and Sabine Flach.
p. cm.
Includes index.
ISBN 978-1-4094-4870-9 (hardcover : alk. paper)
1. Aesthetics. 2. Darwin, Charles, 1809–1882. I. Larson, Barbara (Barbara Jean) II. Flach, Sabine.

BH39.D375 2012
111'.85—dc23

2012012832

ISBN 13: 978-1-138-25267-7 (pbk)
ISBN 13: 978-1-4094-4870-9 (hbk)

DARWIN AND THEORIES OF AESTHETICS AND CULTURAL HISTORY

Darwin and Theories of Aesthetics and Cultural History is a significant contribution to the fields of theory, Darwin studies, and cultural history. This collection of eight essays is the first volume to address, from the point of view of art and literary historians, Darwin's intersections with aesthetic theories and cultural histories from the eighteenth century to the present day. Among the philosophers of art influenced by Darwinian evolution and considered in this collection are Alois Riegl, Ruskin, and Aby Warburg. This stimulating collection ranges in content from essays on the influence of eighteenth-century aesthetic theory on Darwin and nineteenth-century debates circulating around beauty to the study of evolutionary models in contemporary art.

Barbara Larson is Associate Professor of Art History at the University of West Florida. She specializes in intersections between art and science in the nineteenth and twentieth centuries. Her publications include The Dark Side of Nature: Science, Society, and the Fantastic in the Work of Odilon Redon *(2005), and* The Art of Evolution: Darwin, Darwinisms, and Visual Culture *(2009), which she co-edited.*

Sabine Flach is currently Professor of Contemporary Art and Art Theory at the School of Visual Arts, Department of Fine Arts, New York City. She specializes in modern and contemporary art, aesthetics, aesthesis, and media of embodiment; habitus in habitat; the epistemology and aesthetics of visual thinking; and intersections between art, art theory, art history, and cognitive, neuro- life-sciences, and perceptual psychology. Her publications include: Habitus in Habitat I: Emotion and Motion *(2010),* Habitus in Habitat II: Other Sides of Cognition *(2010), and* Habitus in Habitat III: Synaesthesia and Kinaesthetics *(2011), which she co-edited, as well as* Sensing Senses. Die WissensKünste der Avantgarden. Künstlerische Theorie und Praxis zwischen Wahrnehmungswissenschaft, Kunst und Medien. 1915–1930 *(2013).*

This book is dedicated to the memory of Chester A. Larson.

Contents

List of illustrations

Notes on contributors

Sabine Flach is currently Professor of Contemporary Art and Art Theory at the School of Visual Arts, Department of Fine Arts, New York City. She specializes in modern and contemporary art, aesthetics, aesthesis, and media of embodiment; habitus in habitat; the epistemology and aesthetics of visual thinking; and intersections between art, art theory, art history, and cognitive, neuro- life-sciences, and perceptual psychology. Her publications include: *Habitus in Habitat I: Emotion and Motion* (2010), *Habitus in Habitat II: Other Sides of Cognition* (2010), and *Habitus in Habitat III: Synaesthesia and Kinaesthetics* (2011), which she co-edited, as well as *Sensing Senses. Die WissensKünste der Avantgarden. Künstlerische Theorie und Praxis zwischen Wahrnehmungswissenschaft, Kunst und Medien. 1915–1930* (2013).

Barbara Larson is Associate Professor of Art History at the University of West Florida. She specializes in intersections between art and science in the nineteenth and twentieth centuries. Her publications include *The Dark Side of Nature: Science, Society, and the Fantastic in the Work of Odilon Redon* (2005), and *The Art of Evolution: Darwin, Darwinisms, and Visual Culture* (2009), which she co-edited.

Ellen K. Levy, a New York-based artist, is past president of the College Art Association and Special Advisor on the Arts and Sciences at IDSVA. Levy has exhibited her work widely, both in the US and abroad, including at the New York Academy of Sciences and the National Academy of Sciences.

Marsha Morton is a Professor of Art History at Pratt Institute in Brooklyn. She has lectured frequently and published many essays on topics which address art, science, and music, especially in nineteenth-century Germany and Austria. Her books include *The Arts Entwined: Music and Painting in the Nineteenth Century* (2000), which she co-edited, and the forthcoming *Max Klinger and Wilhelmine Culture: On the Threshold of German Modernism.*

LAURENCE SHAFE is a doctoral candidate at the University of Bristol, where he is completing a dissertation on Darwin and beauty.

LARRY SILVER is Farquhar Professor of Art History at the University of Pennsylvania, past president of the College Art Association, and a founder and editor-in-chief of *caa.reviews*, its on-line reviews journal. A specialist in painting and graphics from Germany and the arts of the Netherlands, he has focused on both production and consumption of images during the early centuries of printmaking and the open art market, most directly in *Peasant Scenes and Landscapes* (2006).

JAN SÖFFNER works as a research associate at the Kolleg Morphomata in Cologne and is former research associate at the Zentrum für Literatur- und Kulturforschung, Berlin. His current research focuses on emotion, moods, theories and practices of mimesis, and embodiment in western literature.

SIGRID WEIGEL is director of the Zentrum für Literatur- und Kulturforschung, Berlin. She has published on Heine, Warburg, Freud, Benjamin, Arendt, and the cultural history of science. Recent books include *Heine und Freud. Die Enden der Literatur und die Anfänge der Kulturwissenschaft* (2010), *Walter Benjamin. Die Kreatur, das Heilige, die Bilder* (2008), and *Genea-Logik. Generation, Tradition und Evolution zwischen Natur- und Kulturwissenschaften* (2006).

Preface

This book was inspired by two conferences in which the influence of aesthetic theory on Darwin and the scientist's important legacy to art historians and aesthetic theorists came into focus. The first conference, "Kultur der Evolution: Rethinking Evolutionary Theory from the Perspective of Cultural Studies," was sponsored by the Zentrum für Literatur- und Kulturforschung in Berlin in October 2008, and the second conference, "The Art of Evolution: Darwin and Visual Culture" was held at the Courtauld Institute of Art, University of London, on the occasion of the Darwin bicentennial in July 2009. The very richness of the topic at hand is such that the essays in this volume begin with Darwin's youth and continue into the present. As the reader will gather in the following introduction, there is much more to be investigated concerning Darwin, aesthetic theory, and cultural history, and the editors hope that the text at hand will encourage further enquiry.

Introduction

Barbara Larson

In the late 1830s, when Charles Darwin was formulating his theories of evolution, including the reason behind the existence of the emotions (formerly, the passions) and their manner of expression, eighteenth-century aesthetic theory provided a serious resource. Notebooks from 1838 to 1844 are filled with references to Joshua Reynolds, Edmund Burke, Archibald Alison, and David Hume, among others. Then, as in recent decades, neither scientists, nor philosophers, nor art theorists were thought to have a privileged position in explicating emotive responses to forms and colors in the natural world or in the domain of the arts. Darwin's ideas were shaped by many observers on the human (or animal) response to the natural world, especially, but not exclusively, those engaged in neurology, including a number of aesthetic theorists. Today, an active site of aesthetic theory is emerging from current work in neuroscience, and Darwinian ideas have proven to be central, as we shall see.

While a round through the neurosciences might make sense in terms of a physiological basis for aesthetic theory that we might expect to be connected to Darwin, what of the "in between" period, when other kinds of theories supplanted or were more readily considered than a biological basis for an aesthetic appreciation of nature and art? Indeed, when Darwin was publishing his principle texts, other positions were far better known and more influential than a physiological basis for aesthetics, such as the moralizing of Ruskin on the spiritual underpinning of beauty in nature and, by extension, in the best of art, and the Aesthetic "art for art's sake" Movement—neither moral nor based in biological concerns. Yet a number of theorists of the arts either used Darwinian language, referred to the scientist in crucial aspects of their work, or enthusiastically recalled his writings as central to their own developing ideas. These include philosophers of art as divergent as Hippolyte Taine, Alois Riegl, Gottfried Semper, Heinrich Wölfflin, Aby Warburg, George Santayana, and E.H. Gombrich.

In some cases, cultural histories were implicated in the idea of selective external pressures that "weed out" the unviable, leaving only the environmentally suitable, according to Darwinian natural selection. Devoted Darwinian follower Grant Allen had taken up this approach in *Evolution in Italian Art* (1908). In it, he claimed a limited number of possible subjects available to Italian artists during the Renaissance and a "descent with modification," which he examines "with the eye of an evolutionist." Thus, varieties emerge that reflect different environmental conditions imposed upon them; for example, the "Paduan type," which "befits the denizen of a university city." This "scientific" approach to the history of art itself has a considerable legacy, as we shall see. Moreover, we find Darwin emerging at unlikely aesthetic junctures, such as in the early twentieth century, when artists mined his ideas on camouflage to affect the aesthetic act of the "hidden" object in plain view wherein the end goal is protection from one's enemy, ultimately contributing to the promotion of camouflage during World War I. In another context that involves survival and its antithesis, are corporeal aesthetics of health and the body in eugenics campaigns in which Darwin's ideas on selection were often appropriated and underscored as authoritative.

As a student at Cambridge in the late 1820s, Darwin had read Joshua Reynolds' *Discourses on Art* and Edmund Burke's *A Philosophical Enquiry into the Origin of our Ideas of the Sublime and Beautiful* and had visited the paintings collections at the National Gallery in London and at the Fitzwilliam in Cambridge.[1] As would be the case with a gentleman scholar of Darwin's generation, he was familiar with the by then standard aesthetic categories of the sublime, the beautiful, and the picturesque. Like the artists who accompanied voyages of exploration, he often framed his pictorial references in this familiar context during the course of his five-year voyage around the world.[2] In his *Beagle* diary of the early-to-mid 1830s, and later in the published *Journal of Researches*, novel scenes were often described according to such conventions. On approaching Santa Cruz:

> The gaudy coloured houses of white yellow & red; the oriental-looking Churches & the low dark batteries, with the bright Spanish flag waving over them were all most picturesque.—The small trading vessels with their raking masts & the magnificent back ground of Volcanic rock would together have made a most beautiful picture.[3]

When he refers to several soldiers playing cards by a campfire at night on the Argentine pampas as a "Salvator Rosa scene," he is demonstrating awareness of the British association of the sublime with the example of the seventeenth-century painter.[4] In concluding comments on his voyage, real scenes and artificial representation converge: "Group masses of naked rock, even in the wildest forms; for a time they may afford a sublime spectacle… paint them with bright and varied colours, they will become fantastick; clothe them with vegetation, they must form, at least a decent, if not a most beautiful picture."[5] In his discussion of color, Darwin was evocative and often precise. He relied upon Patrick Syme's hand-tinted *Werner's Nomenclature of Colours,* but sometimes mentioning the art of painting itself was enough.[6] In December of 1831, when approaching the Bay of Biscay, he wrote:

> [I] was much struck by the appearance of the sea... It is not only the darkness of the blue, but the brilliancy of its tint when contrasting with the white curling tip that gives such a novel beauty to the scene.—I have seen paintings that give a faithful idea of it.[7]

After the publication of his experiences and some of his scientific observations aboard the *Beagle,* Darwin turned once again to Reynolds and Burke, and to many other sources as well, as he worked privately on the theories that would come to have such a tremendous impact on science and culture over the next two centuries. Aesthetic theorists now were helpful to him not in "picturing" nature, but in grappling with the physiology behind responses to the natural world. Darwin was drawn to the "sensationalist" school of philosophy of the mind in British eighteenth-century thought that originated with Locke, who connected subjective human response to objective phenomena in the environment. Followers of this school made the vibratory patterns of the nervous system central in understanding the relationship of man to his surroundings. These vibrations, following the impact of the external object on the nervous system, were described as sensations. For Locke, sensations are followed by reflection, which shape the self. Locke believed that the effect of sensations on the *tabula rasa* of the newborn contributed to associations or memories and were crucial to development.

Darwin thought Edmund Burke worthy of consideration. Burke drew on the materialism of Locke, but had less of an interest in associationism. Burke made pleasure and pain the two central arenas of aesthetic impact in his program—the former attached to society (and beauty) and the latter to self-preservation (and the sublime). Though not a scientist himself, he posited that beauty is produced through smoothness and gradual variations in external objects, producing a relaxing effect on the nerves, whereas responses connected to terror and pain (the sublime) cause extreme tension in the nerves. The speculative advantage for Darwin was the universal application of such possibilities, an emphasis on innate responses (rather than those which are learned), and a program dividing the individual under threat, which could be applied to the struggle for life within and without species, countered by a theory of the need for and pleasure found in society.

The Scottish branch of associationism was compelling as well. These thinkers focused on the nervous system and stressed past memories or associations. For Hume, whose *A Treatise of Human Nature* had been published in 1739–1740 and *Of the Standard of Taste* appeared in the same year that Burke initially published *A Philosophical Enquiry* (1757), the source of aesthetics rested within the mind. However, Hume believed that the physical structure of the mind made some objects more readily beautiful and others likely to inspire fear. Archibald Alison, in his *Essays on the Nature and Principles of Taste* (1790) went further in abandoning the idea of aesthetic possibilities residing within the object itself. Aesthetic appreciation was ultimately found in the free play of the imagination rather than a direct impression from object to mind. We have individual associations or memories that cause responses. Following Locke, we find a face beautiful, for example, not because of set laws of beauty, but because of past associations. Darwin's grandfather, the evolutionist Erasmus Darwin, who was attentively read by Charles Darwin in the years he formed his central ideas, followed such associationist thinking:

> Our perception of beauty consists in our recognition by the sense of vision of those objects, first, which have inspired our love by the pleasure they have afforded to many of our senses; as to our sense of warmth, of touch, of smell, of taste, of hunger and thirst; and secondly which bear any analogy of form to such objects.[8]

Writing in the early nineteenth century, the Scotsman Dugald Stewart also found aesthetic associations to be located in the mind rather than in the object, but this occurs through a complex series of associations. Darwin outlined Stewart's theories of the sublime and of taste. In his autobiography Darwin noted that while on the *Beagle* voyage, he had thought sublime sensations were attached to a higher power (god), but had since realized this was not the case (see Chapter 1 of this volume). Stewart's explanation for such confusion, as noted by Darwin, was that we understand the meaning of the sublime to be, at its most literal, height, and we associate ascension with power (or the sensation of "inward glorying") and therefore think of god as living in the heavens. Later, when we experience wonder or terror as in observing a vast ocean, for example, we are reminded of the rather similar sensations we experience with height and this brings god to mind; thus, through a complex interaction of associations we associate god with the sublime. Darwin offered a somewhat Burkean corrective: "It appears to me, that we may often trace the source of the 'inward glorying' to the greatness of an object itself or the ideas excited & associated with it..."[9] However, the Scottish program encouraged him to consider variability and relativity in standards of beauty. Therefore, humans followed cultural standards of beauty in mate choice and even highly personal ideas based on past associations.

But what of Joshua Reynolds and Darwin's readings in the late thirties? Reynolds is hardly a theorist who would be considered as grounded in biology; on the contrary, his program held up classical standards of the ideal. Yet Reynolds did acknowledge relativism and opinion in relationship to beauty, to which Darwin responded, "Is our idea of beauty, that which we have been most generally accustomed to...[d]eduction from this would be that a mountaineer...borne out of country yet would love mountains, & a negro, similarly treated would think negress beautiful..." Darwin further commented that although Reynolds could account for the instinct to feel beauty, he could not account for the feeling itself.[10] Later, in preparing *The Expression of the Emotions in Man and Animals* (1872), he found Reynolds' observations of the human figure worth reviewing and mentions him, perhaps because this book was designed to appeal to artists as well as scientists.[11]

Darwin's own most developed aesthetic program was in the area of sexual selection wherein he believed that females of all species save humans choose the most "beautiful" male, whether this is based on color or formation of secondary sexual characteristics (e.g. antlers) or size or a combination thereof, and human males choose females also based on ideas of beauty. Sexual selection was introduced in *On the Origin of Species* (1859), but largely developed in later years and was a major consideration in *The Descent of Man, and Selection in Relation to Sex* (1871). The impact of the Darwinian program on the aesthetics of allure among humans has been the subject of a number of books in literary theory and more recently has come into focus in the arts.[12] Biologist Geoffrey Miller has suggested that it is sexual choice with all its aesthetic implications that has been the driving force of human evolution (as opposed to the primacy of that other Darwinian mechanism "natural selection" or "survival of the fittest.").[13] This includes abilities in the arts, sports, and leadership that are forms of display and become interwoven with mating rituals (including in the present day).

Darwin's extensive speculations on the biological basis of mate attraction (where beauty effects procreation and survival of the species) or the symbiotic relationship between birds or insects and plants and flowers (where beauty as perceived by humans is but a byproduct of natural processes, such as the ability of those seeking nectar to spot bright colors) drew considerable attention, much of it negative. The eminent art theorist, John Ruskin, who thought of himself as a skilled scientist, was a self-appointed detractor, promoting the idea that beauty in nature was for the aesthetic pleasure of god and man. Jonathan Smith has written about the specifics and timing of Ruskin's publications on plants, for example, designed to thwart Darwin.[14] When the formerly devoted follower of Ruskin, Dante Gabriel Rossetti, seemingly diverged to produce sensual "material" paintings of women compressed in tight spaces beginning in 1859—a date that coincided with the publication of Darwin's *On the Origin of Species*—there were those who suspected some kind of theoretical alignment between the scientist's physiological arguments on evolution and the emerging amoral school of Aestheticism, most notably that branch derisively dubbed the "fleshly school," which included Rossetti, by poet Robert Buchanan.[15]

Though Rossetti never claimed a biological or Darwinian basis to his approach, the theorist who did attempt to popularize the scientist's aesthetic ideas was Grant Allen. In a direct rebuttal to Ruskin and his followers (perhaps Buchanan among them) he opened his text *Physiological Aesthetics* (1877) as follows:

> 'Why we receive pleasure from some forms and not from others,' says professor Ruskin, 'is no more to be asked or answered than why we like sugar and dislike wormwood.' The questions thus summarily dismissed by our great living authority on Aesthetics are exactly the ones that this little book asks and, I hope, answers.[16]

Allen turned to Darwin's interest in pleasure and pain and also credited natural selection with prompting our aesthetic tastes. His aesthetic categories were not the beautiful and the sublime, but rather the beautiful and the ugly (the latter causing disturbing sensations to the nervous system and of less concern than the beautiful). Allen had been influenced by Alexander Bain who based aesthetics largely on pleasure. He found in Bain as well as in Herbert Spencer (*Principles of Psychology*, 1855) the conviction that all mental processes were based on physical sensations rather than spontaneous thought. Like Spencer, he believed in increasing evolutionary complexity within individuals (in a hierarchical sense) and the history of society. Those with "indiscriminate nervous organization, an untrained attention, a low emotional nature, and imperfect intelligence" could not take pleasure in either art or the beauties of nature, the latter being of the greatest importance. By the 1870s, Allen's ideas appeared far more modern than those of Ruskin in certain quarters, and his volume had some impact at that most venerable of institutions, the Royal Academy. Two years later, he published *The Colour-Sense; Its Origin and Development. An Essay in Comparative Psychology*. In this volume, Allen discussed the development of color detection in all creatures. Human ancestors developed an attachment to bright colors through the search for food, and this attachment was shared by all fruit-eating animals and living beings who made use of flowers. Indeed, insects and birds were engaged in a symbiotic relationship with flora, aesthetic beauty resulting from their interdependence. Bright colors were also important in mate selection and in recent times could be found in the arts. Fifteen years later, Allen went further still, ascribing all pleasure and beauty to the sex instinct.[17] From his perspective all "loveliness" in nature was based on Darwinian sexual selection. Even all of humanity's higher emotions, including paternal and maternal love, could be traced to the sex instinct. George Santayana, one of the first to publish on aesthetic theory in the US, assigned Grant Allen's *Physiological Aesthetics* in the first class taught on the subject at Harvard (1892–1893). In his own text based on his course lectures, *The Sense of Beauty* (1896), these words were included in his opening paragraph:

> Of late we have even learned that the forms of many animals are due to the survival by sexual selection of the colors and forms most attractive to the eye. There must therefore be in our nature a very radical and wide-spread tendency to observe beauty and to value it.[18]

Darwin's cousin, Francis Galton, appropriated ideas on natural selection to uphold his eugenic theories, disseminated into culture both in Britain and abroad in light of healthful, fit, beautiful bodies. These perfect and appropriately erotic bodies were then enmeshed in aesthetic theories of sexual selection. While the ominous implications of the wide spread of eugenics has led to a lengthy history of absence of scholarly work on the decades-long pervasiveness of these corporeal and reproductive ideals in visual culture, recently this lacuna has been addressed in the scholarly literature.[19] Barbara Creed has examined the lighter side of the by then wide popular acceptance of evolutionary ideas regarding the body by demonstrating the Darwinian roots of both degenerative and monstrous corporeal types (given the transformative properties of evolution with its dark degenerative potential) and healthful, reproductive, beautiful bodies in early Hollywood films. Creed writes about the Darwinian "mating game" as celebrated in such cinematic productions of the 1930s as Busby Berkeley musicals with their tap-dancing bevies of beauties sporting "fruit" costumes, aerial portrayal of choreographed "flowers blooming," and male wooers penetrating their combined center as if through a pistil, singing such refrains as "I'm young and healthy":

> More than any other art form, the cinema, with its power to transform an image into a spectacle, has responded to Darwin's sexualisation of beauty…The power of the visual image and special-effects technologies… have combined with the cult of stardom to present beauty as erotic, alluring and captivating, leading to a sexualisation of beauty.[20]

The aesthetic implications of the ability of an organism to "disappear" by imitating its surroundings or another living entity is an organic aesthetic "strategy" that fascinated Darwin and his followers. Camouflage refers to the ability of a creature to appear as part of its environment while mimicry is an adaptation to look (or sound) like another species. Oxford biologist Edward Poulton devoted himself to promoting Darwin's natural selection as the overriding mechanism behind mimicry and camouflage at the end of the nineteenth century. His follower, the painter Abbott Thayer, not only used Darwinian ideas on camouflage as the basis for his nature paintings, but through personal observation promoted both artistically and scientifically the importance of the countershading of an animal's coat to survival (as in an animal's white belly), wherein forms appear less solid and round, thereafter called *Thayer's Law*, along with disruptive coloration (as in a zebra's stripes). In 1909, he and his son Gerald Thayer authored the book *Concealing-Coloration in the Animal Kingdom: An Exposition of the Laws of Disguise through Color and Pattern, Being a Summary of Abbott Thayer's Disclosures* (New York: Macmillan). During World War I, Thayer and his followers successfully proposed ideas to camouflage ships, instituted through the American Camouflage Corp.

Despite the deeply ingrained influence of Lamarkian evolution in France, Darwinian ideas began to reach a fairly large audience of artists as early as the 1860s through the theorist Hippolyte Taine and his lectures at the École des Beaux-Arts. Already convinced that the spirit and form of a people reflect the environment and the time period in which they find themselves—an idea captured in his well known dictum "race–milieu–moment"—he had likened this relationship to that between a plant and its species and surroundings. In 1865, in a lecture to his students on the historical moment and artistic forms, published in his famous *Philosophie de l'art,* he gave this advice:

> Physical temperature acts by elimination and suppression—in other words, by *natural selection*. Such is the great law by which we now explain the origin and structure of diverse and existing organisms—a law as applicable to moral and physical conditions, to history as well as to plant and animal. In short, there is a *moral* temperature, consisting of the general state of mind and manners…a moral temperature makes a selection among different species of talent, allowing only this species or that to develop, to the exclusion of others. It is through some such mechanism that you see in artistic schools at certain times the sentiment of the ideal.[21]

That culture might respond to similar pressures found in the biological world would become an integral idea in histories of art and style in Germany. Semper, Riegl, and Wölfflin were all concerned with understanding formal changes over time. Semper identified *Urformen* (prototypical forms) and the subsequent development of form through environmental conditions and circumstances, e.g. a formula akin to "descent with modification," no doubt inspired by his reading of Darwin's *Origin of Species,* despite an antipathy to determinism and what it might suggest concerning free will.[22] The architectural theorist Georg Heuser explicitly conjoined ideas from Semper and Darwin to promote a technological history of architecture with natural selection as the motivating force.[23]

Similarly, Conrad Fiedler also applied Darwinian language to architectural history in *Observation on the Nature and History of Architecture* (1878). Riegl also followed a "descent with modification" schema in which formerly conceived periods of decay are accompanied by the budding of new life and new forms appropriate to the time. He used the term *Kunstwollen* to suggest the new "adaptations" of inherited form. Sensory experience was important to Riegl as well, and he was interested in the work of physiologist Sigmund Exner, part of a circle working in Austria in the late nineteenth century attempting to understand the impact of evolution on aesthetics. Exner emphasized the evolution of the nervous system and a parallel processing in the brain of memory traces and sensory stimuli in cultural adaptation. Sensory stimuli causing pleasure would then be associated with beauty and this would often have a cultural basis.[24] Like Riegl, Wölfflin also used the analogy of the "bud" of a new style with an inherent basis in old forms in his *Principles of Art History* (1915). Later in the twentieth century, E.H. Gombrich found cultural applications of Darwin's selection worthy of consideration.[25]

Darwin's *Expression of the Emotions* influenced the awareness of the emotional power of art and primitive sources in expression in Germany, where it had a significant impact. As we have seen in Darwin's study of eighteenth-century aesthetic philosophy, he was interested in the physiological basis of all emotions. Human physiology was recognized by Robert Vischer as the basis for aesthetic response, exemplified by his concept of "empathy" formulated in 1873, one year after the publication of *Expression*. As in the eighteenth century, "nerve vibrations" were responsible. Through a visual and general corporeal experience, nerve vibrations force an object or experience into focus and then it enters the conscious mind, resulting in a symbolic picture. Touch mattered as well as sight and the overall experience is an empathetic one.[26] For Vischer, our drive to "preserve ourselves" is part of our emotional response which also includes the drive towards well being.[27] In the emotional feedback loop from object to subject to anthropomorphic envisioning of the object, the response is first immediate and sensory, then responsive and kinesthetic.[28] Wölfflin addressed Vischer's ideas by positing that our psychological tendency to project our own bodies onto the objects we view "animates" them, producing kinesthesia.[29]

Aby Warburg enthused later in life about his original reading of *The Expression of the Emotions* in 1888 and how this text thereafter was central to his ideas. The expressive emotional force of the individual illustrations (photographs and wood engravings) captured his imagination. He saw in their extreme gestures and aroused states an encounter between the individual and the chaotic world in which the struggle for existence is found in heightened responses. In outlining the importance of Darwin's book for Warburg, E.H. Gombrich noted that Warburg was steeped in the popular notion that a given historical period could have a means of corporeal expression or mental experience, a spirit of the age.[30] Warburg believed that evolution was teleological with an increasingly advanced state of mental activity and therefore through time there was an inevitable suppression of frenzied gestures; thus one could trace a "historical psychology of human expression." However, emotive corporeal forms with their "Dionysian" qualities—pathos formula—lie latent in cultural memory.

Unlike Warburg's fascination for regressive, "irrational" responses in human nature to the environment, around the same time in Germany, Darwinian follower Ernst Haeckel promoted an understanding of the human capacity to accept a lyrical form of beauty through an appreciation of the symmetrical forms (in all their intricacy) underlying natural entities, and this had a significant impact on design. Recently, Stephen Eisenman has argued that Darwin's botanical ideas had an influence on design aesthetics in England such as in the work of Christopher Dresser.[31]

In recent years, Darwin and evolutionary aspects of aesthetic theory have come into focus again, though the theorists are largely from the fields of anthropology and psychology. Titles of publications along these lines include *Evolutionary Aesthetics* (Eckart Voland and Karl Grammer, eds, Berlin:

Springer Verlag, 2003), *Evolutionary and Neurocognitive Approaches to Aesthetics, Creativity and the Arts* (Colin Martindale, Paul Locher and Vladimir Petrov, eds, Amityville: Baywood Publishing, 2007), and Denis Dutton's *The Art Instinct: Beauty, Pleasure, and Human Evolution* (New York: Bloomsbury Press, 2009). Evolutionary psychology accepts Darwin's natural selection and extends it to the psyche, positing that our mental capacities are themselves adaptations laid down through natural and sexual selection. Through generations of ancestral experiences we are not born a *tabula rasa*, aesthetically speaking, but have innate tastes. As psychologist Randy Thornhill puts it in the first chapter of *Evolutionary Aesthetics*,

> Beauty is a promise of function in the environments in which humans evolved, i.e., of high likelihood of survival and reproductive success in the environments of human evolutionary history. Ugliness (which evolutionary psychologists seem to prefer as a counter-thesis rather than the sublime) is the promise of low survival and reproductive failure.[32]

An example of how evolutionary psychologists have applied this is the recent popular notion that humans respond aesthetically in a most positive fashion to "savannahs" (in nature and by extension in landscape art) because our ancestors had found this kind of habitat hospitable (and we like branching trees because it reminds us of climbing and escape from danger). It should be noted that not all evolutionary psychologists agree on all points; we may prefer other kinds of habitats (and landscapes in art) because of personal familiarity.

Thornhill notes how psychologists are not all in agreement in terms of corporeal aesthetics either. Alfred Wallace (who shared the discovery of natural selection) diverged from Darwin in explicating a theory that "extravagant features" were to signal phenotypic quality as opposed to mere beauty. While all evolutionary psychologists accept sexual selection, according to Thornhill the consensus at present comes down on the side of Wallace.[33] Another category of interest has to do with neural responses to shapes, colors, and lines. For example, Richard Coss has credited natural selection with "pre-attentive" visual pattern recognition such as diamond patterns as "snake skins." Natural selection, according to Coss, also allows us to detect and quickly finish in our minds partial images. Responding to ideas concerning features and survival is neurologist Joseph LeDoux, who has pointed out that an emotional "quick and dirty" response to gross features such as an S-curve form (which passes directly through the thalamus) allows us to leap out of harm's way, before slightly later processing in the visual cortex and the amygdala (which mediates the emotional response) results in a more fine-tuned assessment of whether we have or have not seen a snake.[34]

Increasingly since the 1950s, neurologists have explored aesthetic implications of neural structures from an evolutionary perspective. For example,

Semir Zeki, along with other brain scientists, has examined aesthetics in light of the transformation in different areas of the brain in evolutionary history and the interaction between attentional and emotional areas in viewing something pleasurable.[35] Anthropologists tend to look at aesthetic production as adaptive and or symbolic. For example, Steven Mithen has suggested hand axes are "aesthetic," given their importance in social and sexual contexts. Anthropologist Ellen Dissanayake has argued that aesthetics—heightened expressions—served to separate off special objects and performances. Deemed more attractive and pleasurable, these sites served to bring about social bonding.[36]

Philosopher and aesthetic theorist Wolfgang Welsch has argued that the importance of animal aesthetics and its understanding to human aesthetics has been largely ignored, despite its significance to Darwin. He notes that Darwin suggested the aesthetic possibilities of odors, songs, "antics," and especially "ornaments of many kinds [such as] the most brilliant tints, combs and wattles, beautiful plumes, elongated feathers, top-knots, and so forth" and quotes the scientist on aesthetic similarities between humans and animals (for example, Darwin had written "the same colors and the same sounds are admired by us and by many of the lower animals," "birds […] have nearly the same taste for the beautiful that we do," and "'the high standard of taste' [in animals] generally coincides without our own standard)."[37] While noting Darwin's comments on pre-aesthetic beauty (e.g. simple animals like corals "are ornamented with the most brilliant tints, or are shaded and striped in an elegant manner [as] the direct result of either of chemical nature or the minute structure of their tissues") and proto-aesthetic beauty (e.g. colors and forms of flowers and fruits as bearing a relational structure to insects and birds necessary for pollination), basing his argument on Darwin, Welsch sees true aesthetics (with human implications) as originating in the non-human world with female choice through their "sense of beauty" and involvement in courtship rituals. Welsch counters the tendency of aesthetic theorists in psychology to emphasize utility and survival in the environment, since, as Darwin points out, aesthetic considerations in sexual selection often undermine survival (as in the peacock's tail, which attracts the attention of predators). Arguing that animals experience aesthetics (as gradations of pleasure) motivated by the sex drive, Welsch addresses the idea that humans are not necessarily so rudimentary in aesthetic response (though he cites Freud on sexual sublimation as a general pervasive condition). His answer to the animal / human aesthetic dilemma lies in the neurological system. Here, too, he invokes Darwin and quotes him on aesthetics: "There must be some fundamental cause in the nervous system."[38] Aesthetics is found in the complex neural relationships that have to do with the emotions and intellect. Though some aspects of aesthetics are cultural or of "high taste," Darwin invokes evolutionary history:

> Everyone who admits the principle of evolution [should] reflect that in each member of the vertebrate series the nerve-cells of the brain are the direct offshoots of those possessed by the common progenitor of the whole group. It thus becomes intelligible that the brain and mental faculties should be capable under similar conditions of nearly the same course of development, and consequently of performing nearly the same functions.[39]

We have already seen that Darwin was aware of theories of the nervous system in aesthetics that date back to the eighteenth century, and Erasmus Darwin had stressed continuities between animals and humans and believed that thought or emotion was based on the movement of "neural fibers."

A cognitive–evolutionary perspective with a Darwinian epistemology has developed in recent decades within literary and film criticism as applied to culture studies, and this can address histories of culture and aesthetic issues. Lisa Zunshine traces this evolutionary orientation in culture studies to Raymond Williams' influential publication *The Long Revolution* of 1961.[40] According to Zunshine, art is "a form of communication predicated on the living organism's need to adapt to its constantly changing environment or to find a way to modify that environment...given that our species' environment is first and foremost *other minds*."[41]

This volume is written by art and literary historians and artists who address Darwin's intersections with aesthetic theories and cultural histories from the eighteenth century to the present day. This writer examines Darwin's interest in Burke's interpretation of the sublime for its physiological dimensions and emphasis on self-preservation at a time when the scientist's theories of survival and natural selection crystallized. Burke was of interest in cultural and scientific circles around Darwin's evolutionist grandfather Erasmus Darwin, whose treatise *Zoonomia*, inflected with certain Burkean ideas, was a point of departure for Charles Darwin.

Laurence Shafe examines competing theories of beauty in light of Darwin's biologically-based aesthetic program of sexual selection, including those of Ruskin, Darwin's follower and aesthetic theorist Grant Allen, and the artists of the Aesthetic Movement. In so doing, Shafe uses the symbol of the peacock's tail which was repeatedly invoked by Ruskin, Darwin, and the aesthetes. Marsha Morton examines the art historical program of Alois Riegl, whose thoughts were rooted both in Darwinian ideas of competition in nature and those of Ernst Haeckel (Darwin's German follower), who emphasized order, community, pantheism, and unity. While Riegl's general historical program as "descent with modification," exemplified by his concept of the driving force of *Kunstwollen* as inheritance and adaptation has been previously noted (see above), this chapter argues that his central concepts of haptics and optics are based on shifting strategies for society's relationship with nature. In the ancient world, haptics or "near view" (isolated self-sufficient objects denoting power) existed at a time when the "right of the strongest" prevailed amidst a chaotic environment with implications of "survival of the fittest," whereas in the modern age, individuals are cognizant of nature's violence and their own unprivileged status and the optical or "distant view" (interconnectedness of forms in space) prevails as a utopian ideal of interdependence in nature along the lines of Haeckel's "world spirit."

Cultural selection as theorized by Grant Allen or Riegl, among others, as an applied Darwinian historical model of analysis fell out of favor in the twentieth century with the exception of Gombrich and, to some extent, in the "prime object" narrative proposed in George Kubler's *The Shape of Time*, 1962 (see Chapter 4). The "evolution of art," on the other hand, to convey a general notion of progress with the implied (non-Darwinian) teleology of improvement, itself rooted in the Victorian period, was widespread in the first part of the twentieth century. Thus, early twentieth-century abstraction, for example, was generally thought of as part of the inevitable evolution of art from Symbolism or late Impressionism. Recently, as scholarship has returned to the pros and cons of a Darwinian framework to understand transitions in cultural forms, art historians have shown renewed interest in the idea of selection. In 2007, Robert Bork called for a re-evaluation of the application of the Darwinian model of descent to the field of art history.[42] And in 2008, the panel "Is Evolutionism Still Dead?" at the College Art Association conference opened up the issue of Darwinian applications once again. Larry Silver returns to Darwin and proposes that the Darwinian model does have something to offer. In his essay "Culture and the Shape of Time" (Chapter 4), he examines the case of Antwerp in the sixteenth century. He argues that pictorial genres such as landscapes and scenes of everyday life which emerged for the first time in the context of easel painting, can be followed in their shifts of form and meaning though time in the context of market and geographical conditions—in this case involving cultural isolationism.

Sigrid Weigel investigates the epistemological problems associated with applying evolutionary units to culture, suggesting that we need to consider the methods by which meaning is generated within different frameworks of reference. She also notes the tension between the concept of evolution and that of system when evolution is applied to society and culture. She examines fields in which evolution is commonly applied today (beyond the biological), such as communication, the "evolution of culture" by early homo sapiens, the use of tools and their impact on culture, and of greatest interest today, systems theory. She considers current biological interest in a fusion of natural systems with the idea of self-regulation, and she also returns to the development of evolutionary thinking, finding a cultural legacy in the work of Lamarck that then influences Darwin.

Sabine Flach turns to the interesting case of Aby Warburg, who never wavered from claiming the importance of Darwin to his aesthetic and cultural program. Through Warburg's personal insights, Flach suggests that Darwin's *Expression of the Emotions* with its dramatic imagery can be understood as a form of cultural history or rather of nature (science) and culture in tension. She argues that there is an oscillation between the two in Darwin's images, just as there is between the fact of science and the fiction of the sometimes flamboyant or posed corporeal gesture (and their suggestion of events to the viewer), and the combination of heredity (in expression), habits of gestures, and the current environment in which living figures are located. Jan Söffner examines reasons why classic discussions of semiotics and mimicry are ill-founded by referring to Darwinian mimicry. In so doing, he turns to Surrealist imagery and theory and the "riddles" found there in representation and resemblance and unsettled boundaries between one object or another. He argues that true mimicry itself is phenomenological, environmental, and sensorial, not merely visual, and that this is the case in the Surrealist examples (painting, texts, film) that he explores.

Artist and scientist Ellen K. Levy examines divergent scientific views on natural selection today and how these have influenced the aesthetic choices made by some artists. For example, two major trends in evolutionary thinking currently include on the one hand, the more classic view of many evolutionary psychologists, who posit that the environment works on heritable variations, which is also held by artists using a genetic algorithm model (GA) that mimics the process of natural selection, and on the other hand, the perspective of developmental biologists, who argue for the overriding importance of natural constraints on the developing organism that channel evolution and limit adaptation. This, in turn, is echoed in the work of artists who choose a cellular automaton model (CA) that involves generating complex patterns from simple rules. Models of the brain also reflect different views: evolutionary psychology suggests that cognitive plasticity in humans derives from the interplay of diverse instincts in competition. Another faction stresses the roles of information processing and feedback as shaping cognitive development. These views are also found in some current art production. While focusing on artists, Levy reflects upon the "aesthetic" choices of scientists in presenting their ideas.

Notes

1. The Reynolds is noted in Frederick Burkhardt, Sydney Smith, David Kohn, and William Montgomery, eds, *The Correspondence of Charles Darwin*, 1 (Cambridge: Cambridge University Press, 1985), 71, 95, 397, 428. On the Burke, see Chapter 1 of this volume, endnote 4.
2. On ways in which artists recorded unfamiliar sites according to their training in landscape painting, see the writings of Bernard Smith, especially *European Vision and the South Pacific: 1768–1850* (New Haven, CT: Yale University Press, 1959; reprinted in 1985).
3. R.D. Keynes, ed., *Charles Darwin's* Beagle *Diary* (Cambridge: Cambridge University Press, 2001), 20.

4. On the legacy of Rosa as initiating the sublime in landscape painting, see Charlotte Klonk, *Science and the Perception of Nature: British Landscape Art in the Late Eighteenth and Early Nineteenth Centuries* (New Haven, CT: Yale University Press, 1996), 9–10.

5. Keynes, *Charles Darwin's* Beagle *Diary*, 443.

6. The full title is *Werner's Nomenclature of Colours, with Additions, Arranged So as to Render it Highly Useful to the Arts and Sciences, Particularly Zoology, Botany, Chemistry, Mineralogy and Morbid Anatomy. Annexed to Which are Examples Selected from Well-known Objects in the Animal, Vegetable, and Mineral Kingdoms.* Darwin's personal copy, now in the Darwin Library at Cambridge University, is the second edition of 1821.

7. Keynes, *Charles Darwin's* Beagle *Diary*, 18.

8. *Zoonomia*, 3rd edn (London: J. Johnson, 1802), 1, 145.

9. P.H. Barrett, P.J. Gautrey, S. Herbert, D. Kohn, and S. Smith, transcribers and eds, *Charles Darwin's Notebooks, 1836–1844 : Geology, Transmutation of Species, Metaphysical Enquiries* (Ithaca, NY: Cornell University Press, 1987), 605.

10. See *Notebooks*, 527, 570.

11. Jonathan Smith has noted that Darwin had suggested to his publisher that the book be sent to art journals for review. See *Charles Darwin and Victorian Visual Culture* (Cambridge: Cambridge University Press, 2006), 184. When writing about similarities between crying and laughing, Darwin quotes Reynolds: "It is curious to observe, and it is certainly true, that the extremes of contrary passions are, with very little variation, expressed by the same action." *The Expression of the Emotions in Man and Animals*, ed. Paul Eckman (Oxford: Oxford University Press, 1998), 207.

12. In literature, see Gillian Beer, *Darwin's Plots: Evolutionary Narrative in Darwin, George Eliot and Nineteenth-Century Fiction* (London: Routledge & Kegan Paul, 1983); Richard Kaye, *The Flirt's Tragedy: Desire without End in Victorian and Edwardian Fiction* (Charlottesville, VA: University of Virginia Press, 2002); and George Levine, *Darwin Loves You: Natural Selection and the Re-enchantment of the World* (Princeton, NJ: Princeton University Press, 2006). In art, see Jane Munro, "'More Like a Work of Art than Nature': Darwin, Beauty and Sexual Selection," in *Endless Forms: Charles Darwin, Natural Science, and the Visual Arts*, eds Diana Donald and Jane Munro, exh. cat. (Cambridge: Fitzwilliam Museum, New Haven, CT: Yale Center for British Art), 253–91 and Barbara Larson, "Darwin's Sexual Selection and the Jealous Male in Fin-de-Siècle Art," in *The Art of Evolution: Darwin, Darwinisms, and Visual Culture*, eds Barbara Larson and Fae Brauer (Hanover, NH: University Press of New England, 2009), 173–93.

13. Geoffrey Miller, *The Mating Mind: How Sexual Choice Shaped the Evolution of Human Nature* (London: Heinemann, 2000).

14. See his essay, "Evolutionary Aesthetics and Victorian Visual Culture," in *Endless Forms*, 237–52.

15. See Gowan Dawson, *Darwin, Literature and Victorian Respectability* (Cambridge: Cambridge University Press, 2007), 14–21.

16. Grant Allen, *Physiological Aesthetics* (London: Henry S. King, 1877), vii.

17. Grant Allen, "The New Hedonism," *Fortnightly Review* (March 1894), 377–92. Joy Harvey has noted that by the 1870s in France, some anthropologists identified sexual selection as the basis of beauty among all living things. See Joy Harvey, "Evolutionism Transformed: Positivists and Materialists in the Société d'Anthropologie de Paris from Second Empire to Third Republic," in *The Wider Domain of Evolutionary Thought*, eds D.R. Oldroyd and K. Langham (Dordrecht: D. Reidel, 1983), 289–310.

18. George Santayana, *The Sense of Beauty: Being the Outlines of Aesthetic Theory* (New York: Scribner's Sons, 1896), 1.

19. Fae Brauer and Anthea Callen's volume, *Art, Sex, and Eugenics: Corpus Delecti* (Burlington, VT: Ashgate, 2008) is the first text to address this visual culture and the Darwinian roots of Galton's program. Although Darwin himself was not a eugenist, Galtonian language did enter his later work. See Fae Brauer, "Framing Darwin: A Portrait of Eugenics," in *The Art of Evolution: Darwin, Darwinisms, and Visual Culture*, 144–5.

20. Barbara Creed, *Darwin's Screens: Evolutionary Aesthetics, Time and Sexual Display in the Cinema* (Melbourne: Melbourne University Press, 2009), 111.

21. Hippolyte Taine, *The Philosophy of Art* (trans. John Durand from *Philosophie de l'art*, Paris 1865) in *Lectures on Art*, 1st ser. (New York, 1875), 93–4. Slightly earlier, in his *History of English Literature*

(1863–1864), he directly referred to *Origin*. For a discussion on Taine and evolutionism, see Thomas Munro, *Evolution in the Arts, and Other Theories of Culture History* (Cleveland, OH: The Cleveland Museum of Arts, n.d.), 105–17.

22. P. Steadman, *The Evolution of Designs: Biological Analogy in Architecture and the Applied Arts* (Cambridge: Cambridge University Press, 1979), 73, and W. Kleinbauer, *Modern Perspectives in Western Art History, an Anthology of 20th-Century Writings on the Visual Arts* (Toronto: University Press of Toronto, 1989), 20.

23. See H. Mallgrave, *Gottfried Semper: Architect of the Nineteenth Century* (New Haven, CT: Yale University Press, 1996), 364.

24. On Exner, see Olaf Breidbach, "The Beauties and the Beautiful—Some Considerations from the Perspective of Neuronal Aesthetics," in *Evolutionary Aesthetics*, eds E. Voland and K. Grammer (New York: Springer, 2003), 39–68.

25. See, for example, E.H. Gombrich, "Evolution in the Arts: The Altar Painting, its Ancestry and Progeny," in *Evolution and its Influence*, ed. Allen Grafen (Oxford: Clarendon Press, 1989), 107–25; E.H. Gombrich, *In Search of Cultural History* (Oxford: Clarendon Press, 1969); and John Onians, "Gombrich and Biology," in *E.H. Gombrich in Memoriam*, ed. Paula Lizzarga (Pamplona: Ediciones Universidad de Navarra, 2003), 95–119.

26. See Robert Vischer, "On the Optical Sense of Form: A Contribution to Aesthetics" (1878), in *Empathy, Form, and Space, Problems in German Aesthetics, 1873–1893*, eds H. Mallgrave and E. Ikonomou (Santa Monica, CA: Getty Center Publications, 1994), 89–123.

27. Ibid., 103.

28. Ibid., 92.

29. See H. Wölfflin, "Prolegomena to a Psychology of Architecture" (1886), in *Empathy, Form, and Space Problems in German Aesthetics 1873–1893*, eds H. Mallgrave and E. Ikonomou, 149–90.

30. E.H. Gombrich, "Aby Warburg: His Aims and Methods: An Anniversary Lecture," *Journal of the Warburg and Courtauld Institutes* 62 (1999), 272.

31. Stephen Eisenman, ed., *Design in the Age of Darwin: From William Morris to Frank Lloyd Wright*, exh. cat. (Evanston, IL: Northwestern University, 2008).

32. Thornhill in *Evolutionary Aesthetics*, 9–10.

33. Ibid., 11.

34. Joseph LeDoux, *The Emotional Brain: The Mysterious Underpinnings of Emotional Life* (London: Phoenix, 1999), 166.

35. See Marcos Nadal, Miquel Capó, Enric Munar, Gisèle Marty, and Camilo José Cela-Conde, "Constraining Hypotheses on the Evolution of Art and Aesthetic Appreciation," in *Neuroaesthetics*, eds Martin Skov and Oshin Vartanian (Amityville, NY: Baywood Publishers, 2007), 103–30, and Semir Zeki, "Neural Correlates of Beauty," *Journal of Neurophysiology* 91, 1 (2004), 699–705.

36. Ellen Dissanayake, *Homo Aestheticus* (Seattle, WA: University of Washington Press, 1995).

37. Quoted in Wolfgang Welsch, "Animal Aesthetics," in *Contemporay Aesthetics* 2 (2004), 13.

38. Darwin quoted from *Origin* in ibid.

39. Darwin quoted from *Descent* in ibid.

40. Lisa Zunshine, ed., *Introduction to Cognitive Cultural Studies* (Baltimore, MD: Johns Hopkins Press, 2010), 5.

41. Lisa Zunshine in ibid., 10.

42. Robert Bork, "Art, Science, and Evolution," in *Making Art History: A Changing Discipline and its Institutions*, ed. Elizabeth Mansfield (New York: Routledge, 2007), 187–202.

1

Darwin, Burke, and the biological sublime

Barbara Larson

That the specific term "sublime" occurs numerous times in Darwin's writings aboard the *Beagle* and later when writing the transmutation notebooks and the *Journal of Researches*, demonstrates his general familiarity with a Romantic aesthetic category that had become commonplace by the early nineteenth century. Related terms that Darwin often used in his youth like solitude, gloom, grandeur, savage, irregular, and wildness were part of the rhetoric of the natural sublime. Several scholars have acknowledged Darwin's reference to the sublime in his early years: James Paradis, for example, in discussing the importance of the English literary tradition and David Kohn, Phillip Sloan, and Robert Richards emphasizing Germanic sources in his writings, pointing out Darwin's love of Humboldt, whose sublime voice had roots in Goethe and Kant.[1]

It is true that Darwin himself listed English poets of the so-called excursion tradition like Coleridge and Wordsworth as among his favorites, claiming to have read Wordsworth's "The Wanderer" twice, for example.[2] And he spoke with virtual rapture about Humboldt's personal narrative of his travels in South America, several volumes of which accompanied him on board the *Beagle*.[3] However, the specific influence of Edmund Burke's concept of the sublime in nature developed in his *A Philosophical Enquiry into the Origin of Our Ideas of the Sublime and Beautiful* in 1757 and its relevance to Darwin's developing ideas on evolution has not been much discussed.

Burke's theories were of interest in intellectual circles around Darwin's evolutionist grandfather Erasmus Darwin and his Wedgwood cousins. Charles Darwin originally read Burke in 1828 while at Cambridge University and returned to the slim volume on aesthetics ten years later, after the *Beagle* voyage, when mulling over animal behavior and species transformation.[4] His identification with Burkean ideas such as a rapacious nature, the central drive of self-preservation, and the mind and body relationship in the expression of emotions are the subject of this chapter.

The landscape sublime

Darwin counted Milton's *Paradise Lost* as his favorite volume to take on shore when he left the *Beagle* for as yet unknown adventures. This work was virtually canonical among theorists of the sublime and was Burke's own favorite poet to quote from. In his analysis of the great, of magnificence, and of the totality of nature, Burke's immediate predecessor and direct influence, Joseph Addison, had at first advocated familiarity with this text, before turning later in life to the direct observation of nature as affording the "primary pleasure of the imagination."[5] Nevertheless, Milton remained "a perfect master in all the arts of working on the imagination" and clearly Addison's favorite. It may be that in Darwin's efforts to put together the pieces of shifting ecologies around the world he found the Milton helpful because this important source had by now been long discussed as aiding creative thinking on nature. Speculations on grandeur could engage contemplation on the condition of existence.

Sublime pictures mattered to Darwin as well, although these were elusive truths. For example, in trying to describe the Brazilian rainforest, he fell back upon the memory of two prints. Of one of these he wrote to his mentor John Henslow, "I first saw a Tropical forest in all its sublime grandeur.—Nothing, but the reality can give any idea, how wonderful, how magnificent the scene is...Your engraving is exactly true, but underates [sic] rather than exagerates [sic] the luxuriance—I never experienced such intense delight."[6] He also referred to the contemporary works of the artist John Martin, whose images of a fantastic nature were generally accepted as unreal. Of a view of nature along the coast of Brazil, Darwin recorded, "If faithfully represented...a feeling of distrust [would enter the mind of the viewer] as...is the case in some of [John] Martin's views."[7] Whether pictorial representation in the end fell flat or not, Darwin credited a visual source on the sublime in nature read in early childhood—the illustrated *The Hundred Wonders of the World* by C.C. Clarke (Sir Richard Phillips)—in planting the seed that ultimately led to the *Beagle* voyage. His early years were those in which the landscape sublime in visual culture had reached its apex, with many remarkable views to be seen at the Royal Academy exhibitions.

In Addison's emphasis on the primary pleasures of the imagination through direct visual evidence, he identified god as Author of nature. This notion characterized much of eighteenth- and nineteenth-century understanding of a spectacular and sublime nature. Darwin at first struggled with that tradition. In his autobiography, he discussed the notion of deity and creation, then stated,

> Formerly, I was led by feelings such as those just referred to [on religion] to the firm conviction of the existence of God, and the immortality of the soul. In my Journal I wrote that whilst standing in the midst of the grandeur of a Brazilian forest, it is not possible to give an adequate idea of the higher feelings of wonder, admiration,

and devotion which fill and elevate the mind. I well remember my conviction that there is more in man than the breath of his body. But now the grandest scenes would not cause any such convictions and feelings to rise in my mind.[8]

He reminds the viewer that there is no worldwide belief in one god and continues,

The state of mind which grand scenes formerly excited in me, and which was intimately connected with a belief in God, did not essentially differ from that which is often called the sense of sublimity; and however difficult it may be to explain the genesis of this sense, it can hardly be advanced as an argument for the existence of God, any more than the powerful though vague and similar feelings excited by music.[9]

That is not to say that Darwin ever lost his interest in the validity of the sublime as an emotive category, or his awareness of the popular application of the sublime to landscape. In his autobiography, though he felt compelled to reduce the *Beagle* voyage to under six pages because, he said, he had written about it elsewhere, in those few paragraphs he summarized his most crucial memories: "The glories of the vegetation of the Tropics rise before my mind more vividly than anything else—though the sense of sublimity—which the great deserts of Patagonia and the forest-clad mountains of Tierra del Fuego excited in me, has left an indelible impression on my mind."[10] That he carefully considered what constituted sublime sensations is recorded in an incident by his follower George Romanes. In a conversation regarding ancestral habitation in sublime terrain which produces awe, Darwin had recounted that his most sublime feelings came from standing on top of the Cordillera Mountains. But an hour after retiring for the night, he reappeared in his robe and slippers to say that he had been wrong; that instead he felt it even more so in the forests of Brazil. "I am sure now that I felt most sublime in the forests."[11]

Burke and self-preservation

The landscape sublime is perhaps the best-known category of that aesthetic within visual culture, but Burke's ideas as developed in his treatise were applied principally to physiology, and he dwelled on responses of the body one feels when threatened. The application of Burke's ideas on the sublime to landscape was an extrapolation by Romantic Burke enthusiasts and not a central pursuit of Burke himself. As Burke noted in his introduction, he pursued a "diligent examination of our passions in our own breasts; from a careful survey of the properties of things which…influence those passions; and from a sober and attentive investigation of the laws of nature, by which those properties are capable of affecting the body, and thus exciting our passions."[12] Burke was deeply engaged in a psychological and neurophysiological explanation—

the specifics of how sense impressions affect the nervous system; in the case of the sublime this involved "unnatural tension of the nerves."

Burke's ideas developed as part of "the culture of sensibility" in the eighteenth century, when a science of the senses was being formulated and applied to many areas of society, from manners and morality to aesthetics.[13] Although he himself was not a scientist, Burke had a lifelong interest in medicine and was influenced by George Cheyne's theories of nerve tensions and vibrations to explain sensibility, and by his friend and mentor the physician Christopher Nugent, who, interested in pathologies of the nervous systems (and the vibration, pulsation, and oscillation of bodily fluids and solids), espoused therapeutic regimes of exceedingly strong forces (such as through hydrotherapy) applied to overtaxed nervous systems.[14] Despite his focus on threat and danger, for Burke a paradoxical aspect of the sublime was its positive benefits. There was a "healthful" and even a necessary side to the experience of the sublime when direct danger was avoided or at a distance; it strengthened the nervous system and exercised "delicate organs."[15] The sublime was an active principal that countered lethargy.

For Burke, the sublime rested on self-preservation, excited by thoughts of pain and danger. In his section on power, added to the second edition of 1759, Burke wrote:

> besides these things that *directly* suggest the idea of danger…I know of nothing sublime which is not some modification of power. And this branch rises… from terror, the common stock of every thing that is sublime…the idea of pain, in its highest degree, is much stronger than the highest degree of pleasure…. pain is always inflicted by a power in some way superior, because we never submit to pain willingly. So that strength, violence, pain and terror, are ideas that rush in upon the mind together. Look at a man, or any other animal of prodigious strength…the emotion you feel is…rapine and destruction.[16]

Addison had emphasized sight, but Burke believed that all the senses were important. It was his physiological approach to danger and pain, both real and imagined, and the overriding idea of self-preservation that Darwin ultimately sought out after his voyage, rather than the popularized notion of the sublime applied to awe in light of a spectacular landscape.

Burke ventured to compare humans to other species and found physiological and expressive similarities, ideas that would be further developed by Darwin. Burke wrote:

> ...a man in great pain has his teeth set, his eyebrows are violently contracted, his forehead is wrinkled, his eyes are dragged inwards, and rolled with great vehemence, his hair stands on end…the whole fabric totters…Fear or terror, which is an apprehension of pain or death, exhibits exactly the same effects, approaching in violence to those just mentioned in proportion to the nearness of the cause, and the weakness of the subject. This is not so only in the human species, but I have [witnessed this] in dogs, under an apprehension of punishment, that they have writhed in their

bodies, and yelped, and howled, as if they had actually felt the blows...I conclude that pain, and fear, act on the same part of the body...though somewhat differing in degree.[17]

Darwin made similar comparisons between species and, by the time of the publication of *The Expression of the Emotions in Man and Animals* (1872), had developed three principles he felt were at work in the expression of the passions in all living creatures: habits "to gratify some desire or relieve some sensation" that become fixed in a population; oppositional behaviors, for example, submissive postures in dogs in opposition to aggressive ones; and closest to Burke's explanations—actions of the overly stimulated nervous system in times of perceived distress.[18]

Erasmus Darwin, Stubbs, and Fuseli

One artist of the sublime who was almost immediately influenced by Burke's treatise was George Stubbs. From the early 1760s on, he created images of fear and pain in nature. One of his favorite scenes to repeat was that of a traumatized white horse attacked by a lion. It is perhaps of some interest that the first such scene and one of the largest was created for the Marquess of Rockingham (Charles Watson-Wentworth), who appointed none other than Edmund Burke to be his private secretary in 1765 when he became prime minister of Great Britain (Figure 1.1).[19]

Despite Stubbs' unsubstantiated account of actually having seen a horse attacked by a lion in northern Africa, the horse, a creature revered by the British and associated with civilization—agriculture, sports, or even heroic deeds—is hardly expected in an entanglement with a lion. Although the horse in its original state might have sublime dimensions according to Burke, he could also be identified as a symbol of culture "fit for the plough, the road, the draft…[while] the sublime…comes upon us in the gloomy forest, and in the howling wilderness, in the form of the lion, the tiger, the panther, or rhinoceros."[20] Whereas with Milton a wild creature would have implied Satan and sin, the irrational forces of the natural world began to take center stage by the middle of the eighteenth century.

Stubbs was close to the Wedgwood side of Darwin's family and painted Charles Darwin's grandfather Erasmus Darwin's likeness on Wedgwood pottery.[21] Stubbs stayed at the Wedgwood estate from August to October 1780, experimenting with the pottery and painting Wedgwood's family. Among works of these months is a recreation of one of his horse and lion scenes in blue and white jasperware.

Stubbs and Erasmus Darwin, evolutionist, poet of fantastic nature, and medical practitioner, had many naturalist friends in common, such as Sir Joseph Banks. It was around 1780 that Banks took Stubbs to the home of

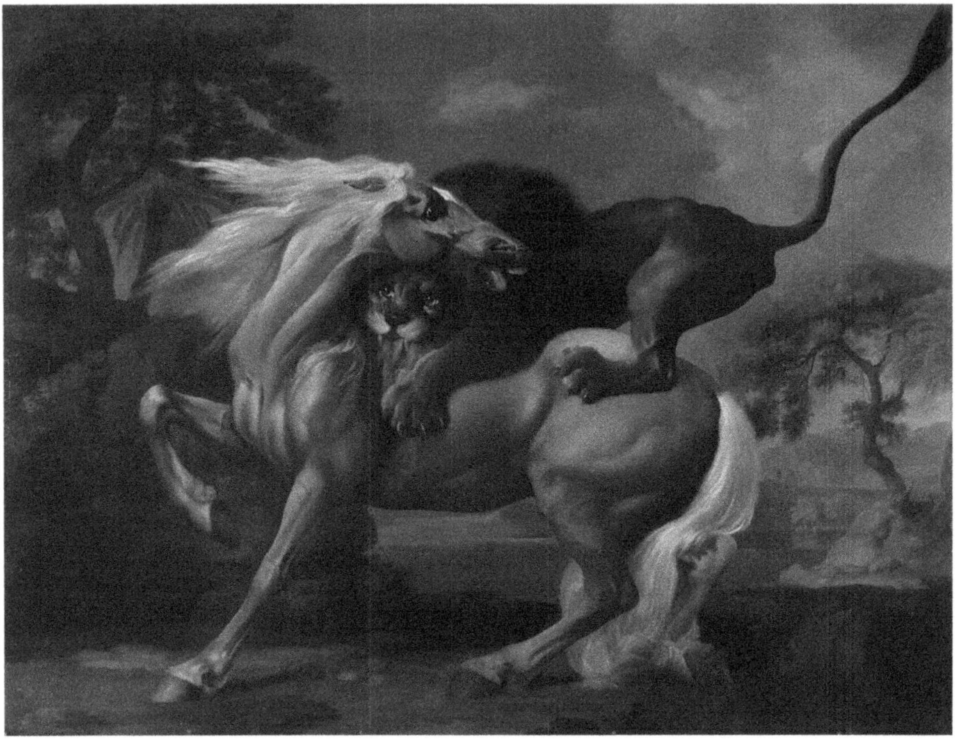

1.1 George Stubbs, *A Lion Attacking a Horse*, 1762. Oil on canvas, 96 × 131 in. (243.7 × 333 cm). Yale Center for British Art, Paul Mellon Collection

Marmaduke Tunstall to draw his tiny "mouse lemur," where he focused on its active digits. Stubbs' exotica extended to moose, rhinoceri, baboons, and cheetahs. His carefully recorded *Portrait of a Monkey* (later identified as a southeast Asian macaque) was first depicted in 1774 and copied again in 1798 (Walker Art Gallery, Liverpool). It sits on a stone while gathering peaches and stops for a moment to regard the viewer nervously, showing the anthropomorphism of the horse. In this case, awareness suggests a simian intelligence that not only foreshadows Charles Darwin's observations of orangutans Tommy and Jenny in Regent's Park several decades later, but demonstrates shared ongoing concerns of the period with Erasmus Darwin, who, at the end of the eighteenth century, was working on his evolutionist poem *The Temple of Nature*, published posthumously in 1803, which included a discussion of human descent. In *The Temple of Nature*, Darwin connected humans with quadrumana. Coleridge and Wordsworth, who in many ways had emulated earlier poetry by Darwin, found this territory for contempt; Coleridge wrote to Wordsworth in 1815 that Darwin's speculations on evolution in terms of orangutans and man was "contrary to religion and possibility."[22] At this time, Stubbs was himself involved in a project of direct comparison between humans and beasts, an ambitious series of anatomical studies of human, tiger, and bird muscular and skeletal systems (1795–1806). He dared to show a human skeleton in a crouching and crawling position, an obvious allusion to similarities with the tiger (Figure 1.2). Included in these drawings is the skeleton of a monkey, upright in a human-like position.[23] Stubbs also created busts of human "passions" in the late eighteenth century, reproduced as a series of prints around 1800; these indicate the dynamic movement of small muscle groups underlying the skin.

Erasmus Darwin befriended another artist who was a reader of Burke, Henry Fuseli. He greatly admired that most sublime of paintings, Fuseli's *The Nightmare* (1781; Figure 1.3).[24] When it was engraved and published in January of 1783 it was accompanied by verses by Darwin that would appear in his *The Botanic Garden*.[25] Both men deeply admired Milton, and Fuseli's dreadful painting with its squat, horrific incubus is reminiscent of Burke's comments in his section on obscurity followed by a quote from Milton. Burke writes, "...night adds to our dread, in all cases of danger, and how much the notions of ghosts and goblins, of which none can form clear ideas, affect minds, which give credit to the popular tales concerning such sorts of beings."[26] Burke then brings up Milton and his brilliant use of obscurity, quotes him, and continues, "In this description all is dark, uncertain, confused, terrible, and sublime to the last degree." Fuseli was a painter of Miltonic subjects, spending much of the decade of the 1790s creating dozens of works inspired by his writings. His Milton Gallery opened in 1799.

While Fuseli's *Nightmare* may seem to have little to do with the sublime from an evolutionary perspective, the smug, monstrous incubus atop its potential victim is not unlike Stubbs' natural, yet equally dangerous, creatures that are out of reach of their victims. It is perhaps of some interest that Erasmus Darwin interpreted the incubus as ape-like. His verses that appeared with the print are: "On her fair bosom sits the demon ape / erect and balances his bloated shape / Rolls in their marble orbs his Gorgon-eyes / And drinks with leathern ears her tender cries." Fuseli's own conviction that apes and humans were much alike must have been another source of discussion between Darwin and Fuseli. Fuseli had defended Jean-Jacques Rousseau's conjecture that orangutans might be semi-human and not a distinct creation and that, moreover, their "society" and very constitutions were not unlike that of humans.[27] Erasmus Darwin was clearly attracted to Fuseli's monstrosities and employed the artist to illustrate his poems of nature's creative and destructive forces in *The Botanic Garden*.

Erasmus Darwin was deeply engaged in the sublime, both in terms of its Romantic cultural associations and in the application of its physiological dimensions. His poetry was often compared to Milton and it included cataclysmic tales of creation, with an explosive universe. In *Economy of Vegetation* (1792),

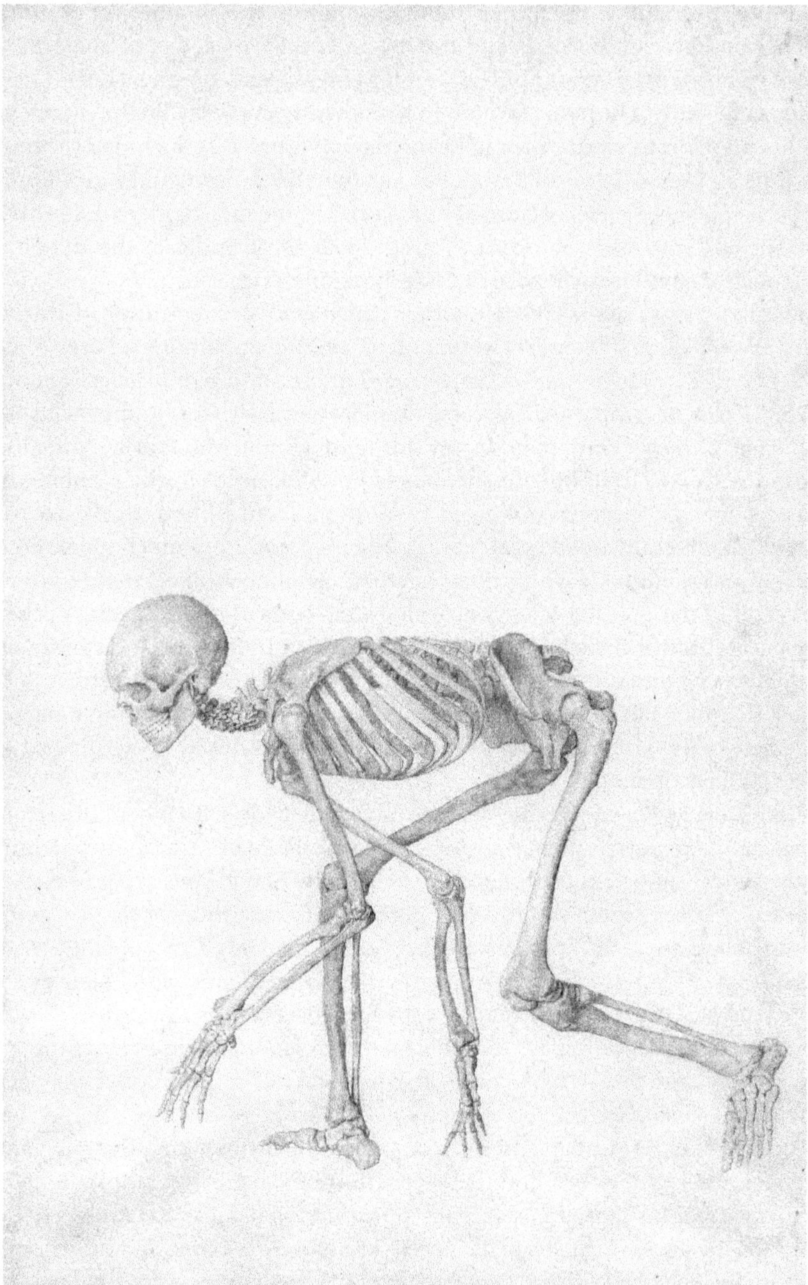

1.2 George Stubbs, *Study of the Skeleton of a Man Stooping*, from *A Comparative and Anatomical Exposition of the Structure of the Human Body with that of a Tiger and a Common Fowl*, c. 1795–1806. Graphite on thin, slightly textured cream wove paper. 17 ⅝ × 11 in. (44.8 × 27.9 cm). Yale Center for British Art, Paul Mellon Collection

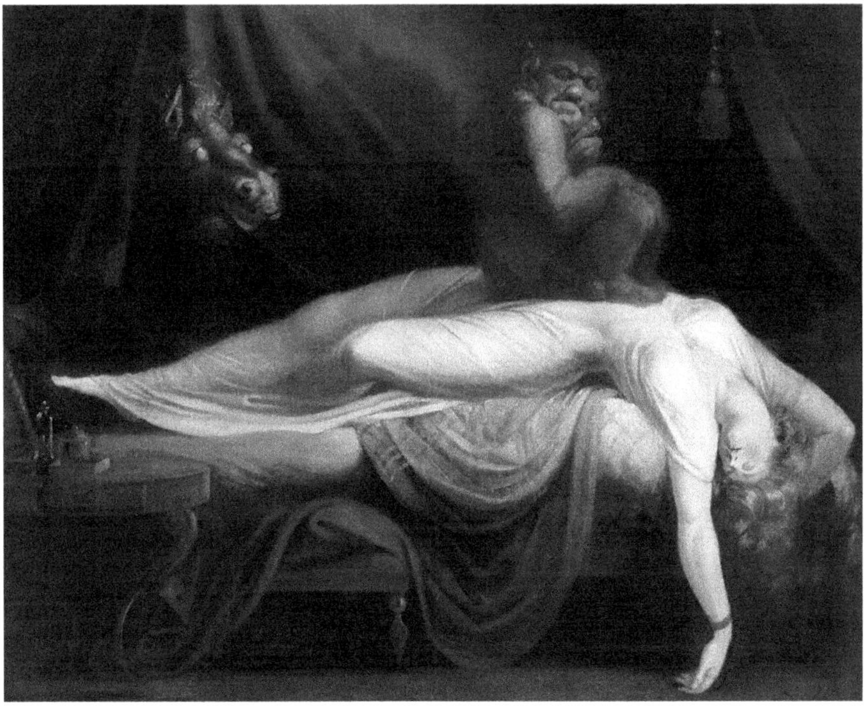

1.3 Henry Fuseli (Füssli, Johann Heinrich) (1741–1825), *The Nightmare*, 1781. Oil on canvas. The Detroit Institute of the Arts, US / Founders society purchase with Mr and Mrs Bert L. Smokler / and Mr and Mrs Lawrence A. Fleischman funds / The Bridgeman Art Library; Swiss copyright

lightning, fireballs, and volcanoes are included in the energies that initiate a spinning world—Horace Walpole referred to Erasmus Darwin's verses as the most sublime he had ever read. But even at his most Miltonic, Darwin thought of himself as based in science. His description of the origins of the universe, which he expanded upon in related notation, was a reference to what today we might refer to as the big bang theory of creation.

Burkean sources in Stubbs and Fuseli that would reverberate in the scientific physiological writings of Erasmus Darwin include the attachment of passions such as dread, fear, and anger to threat or self-protection, but equally close to Burke was Darwin's controversial emphasis on sex or "generation" and its pleasures as a passion leading to survival.[28] Pain and pleasure are as central to Erasmus Darwin's theories as they were to Burke, although he adapted a greater associationist program such as that followed by Locke or Hume, in which attraction and revulsion are learned during the course of one's lifetime through the association of images or ideas arising from sensory input, rather than being innate.

He wrote in his treatise on natural history and medicine, *Zoonomia; or the Laws of Organic Life* (1794):

As the sensations of pleasure and pain are originally introduced by the irritations of external objects: so our desires and aversions are originally introduced by those sensations...when the objects of our pleasures or pains are at a distance; and we cannot instantaneously possess the one, or avoid the other, then desire or aversion is produced, and a voluntary exertion of our ideas or muscles succeeds.[29]

He speculated on adaptive structures that have importance in survival, including those that aided in escape from predators. However, it was the "joy" in sexual reproduction that ensured the continuity as well as the evolutionary potential of the species.[30] His sensual approach to the object of "love" as integrated with notions of beauty in humans was one he shared with Burke. Of a general attraction to smoothness and curves in consideration of beauty, Burke had written:

Observe that part of a beautiful woman where she is perhaps the most beautiful, about the neck and breasts; the smoothness; the softness; the easy and insensible swell; the variety of the surface, which is never for the smallest space the same; the deceitful maze, through which the unsteady eye slides giddily, without knowing where to fix, or whither it is carried.[31]

In his acknowledgement of this passage, Erasmus Darwin responded, "the sentiment of Beauty or Loveliness [is] suggested by easy-flowing curvatures of surface, with smoothness; as is so well illustrated in Mr Burke's Essay on the Sublime and Beautiful."[32] However, for Burke that which we find beautiful may well have been innate to begin with, whereas for Erasmus Darwin it was learned. Darwin continues, "The sentiment of Beauty appears to be attached from our cradles...in the form of the female bosom..."[33]

Erasmus and Charles Darwin and Burke

Erasmus Darwin foreshadowed his grandson in many ways. As an early evolutionist he was originally inspired by fossil evidence. Charles Darwin would later recall how important fossils had been for his own developing views. In Erasmus Darwin's poetry, natural laws over god are paramount. Sexual selection, a mechanism that would be expanded upon by Charles Darwin, with male use of weapons of combat was a mitigating factor. In *Zoonomia*, Erasmus had written that the outcome of rivalry was that "the strongest and most active animal should propagate the species, which should thence be improved."[34] Variation among species was recognized, and there was struggle for survival in a rapacious nature. The origin of even the most complex of living beings came from rudimentary forms. He believed that plants could feel—the basis for his poems *The Loves of the Plants* (1789)—and he shared with Charles Darwin a fascination with that most sublime of plants, the insectivorous *Drosera*. The importance of biological drives like lust and hunger in order to promote survival were observed. Erasmus Darwin seems to have read Malthus' *An Essay on the Principle of Population*, published in 1798, which would prove so instrumental for his grandson in terms of natural selection or "survival of the fittest." In *The Temple of Nature* Darwin wrote, "So human progenies, if unrestrain'd / by climate friended and food sustain'd / o'er seas and soils, prolific hordes! would spread / Erelong, and deluge their terraqueous bed / but war, and pestilence, disease and dearth / sweep the superfluous myriads from the earth." *The Temple of Nature* was originally intended to be entitled *The Origin of Society*, which remains the running head across all the interior pages.

When Charles Darwin opened his first "transmutation of species" notebook (B) in the summer of 1837, he entitled it "Zoonomia."[35] Following Erasmus Darwin, Charles Darwin focused on questions of generation, adaptation, variability, and inheritance. He took extensive notes from his grandfather's publications. He read Lamarck, Lyell, and Cuvier; then he turned to the writings of many other thinkers less engaged in theories of evolution that might prove helpful in his quest to understand animal and human behavior and changes in species. Of these, perhaps the most notable event, even by Darwin's own account, was his reading of Malthus. Darwin read *An Essay on the Principle of Population* in the sixth edition (1826) between September 28 and October 2, 1838, and 40 years later observed of this moment,

> In October 1838, 15 months after I had begun my systematic enquiry, I happened to read for amusement Malthus on *Population*, and being well prepared to appreciate the struggle for existence which everywhere goes on from long-continued observation of the habits of animals and plants, it at once struck me that under these circumstances favorable variations would tend to be preserved, and unfavorable ones to be destroyed. The result of this would be the formation of new species. Here then I had at last got a theory by which to work.[36]

Beginning on October 2, Darwin initiated a new transmutation notebook (E) along with one labeled "Expression" (N) and within the week had mentioned Burke (whom he had been rereading since July), Malthus, and the orangutans Tommy and Jenny from the Zoological Gardens—Jenny with her child-like antics, who plucked hidden ears of corn from straw then brought them to Darwin, and Tommy, who was ill, with an expression of languor and suffering. While Malthus represented on the one hand a solution to speculations on species transformation under the pressure of overpopulation, interspecies struggle, and environmental circumstances which Darwin explored in his transmutation notebooks, on the other hand, he, along with Burke, offered ideas on the implications of the passions in terms of survival that could be considered as Darwin pondered human descent. While he makes no direct reference to his grandfather's "Orang outang theology of the human race" (Coleridge) this could not have been far from his mind at this time or even as early as April 1838, when Darwin had written,

> Let man visit Ourang-Outang in domestication, hear expressive whine, see its intelligence when spoken; as if it understood every word said—see its affection.—to those it knew.—see its passion & rage, sulkiness, & very actions of despair; "let him look at savage, roasting his parent, naked, artless, not improving yet improvable & then let him boast of his proud preeminence…"[37]

It may have been that looking back at his grandfather's work brought Malthus to Darwin's attention along with reminding him of Burke, still fresh in his memory. The specific passage from Burke he cited in early October was one his grandfather had already acknowledged in *Zoonomia* regarding expression, its continued action on the body, and its usefulness to society. Erasmus Darwin had written the following:

> ...when we put ourselves into the attitude that any passion naturally occasions, we soon in some degree acquire that passion; hence when those that scold indulge themselves in loud oaths,and violent actions of the arms, they increase their anger by the mode of expressing themselves; and on the contrary the counterfeited smile of pleasure in disagreeable company soon brings along with it a portion of the reality as is well illustrated by Mr. Burke (Essay on the Sublime and Beautiful)…From this our aptitude to imitation, arises what is generally understood by the word sympathy…[38]

Focusing on the physiological implications of imitation, Charles Darwin noted, "Burke…says on mimicking expression of emotions, he has felt the passions of a face '& mind' sympathetic with internal organs, as actions of the heart."[39]

In Malthus, Darwin found significance in reading about the passions and their usefulness and hindrance to mankind—this ranged from intemperance in eating and drinking leading to disorders of health or multiplying too fast, leading to death from poverty. However, Malthus's beliefs that "natural and moral evils" were instruments of god was not in keeping with Darwin's developing speculations. In Malthus, instant gratification like drinking too much or sex out of wedlock led to degenerate states, and repression of this, what Malthus referred to as "obstacles thrown in the way very early" as in being taught manners and restraint by one's parents, actually produced greater passions—thus, passions and civilization paradoxically work hand in hand. In this way, Malthus maintained a tidy cultivated world. Burke did not provide Darwin with such supernatural stumbling blocks. And, unlike certain moral philosophers he was then reading, such as James MacIntosh or Adam Smith, who followed an associationist program on learned social behavior, Burke's text offered in many places innate explanations which were more in keeping with Charles Darwin's speculations.[40] As Burke himself noted, "Some things must have been originally and naturally agreeable or disagreeable, from which the others derive their...powers; and it would be, I fancy, to little purpose to look for the cause of our passions in association, until we fail of it in the natural properties of things."[41] Burke was interested in the natural workings of the mind beneath learned responses. Although god or "providence" was never completely removed from his program, Burke believed in an innate and uniform organization of mind, a biological basis to behavior.[42] Neither Burke nor Darwin rejected associationism, but the *tabula rasa* theory of behavior of Locke or Erasmus Darwin which relied entirely on learning through experience, with each organism beginning the process afresh was of little appeal to Charles Darwin.[43]

Darwin's first entries on the *Philosophical Enquiry* in his Notebook M are from July 1838.[44] Burke initiated his volume by laying out the development of the human mind and the origins of the expression of the passions. He identified curiosity as the first and simplest emotion; here he was referring to early childhood, essentially a state of wonder over novelties. His discussion is not entirely unlike Darwin's delighted observation in early October that Jenny, when discovering ears of corn, "just like a child not knowing what to do with them, came several times & opened my hand, & put them in—like a child."[45] Burke quickly passed on to "people advanced in life" and the twin passions revolving around pleasure and pain. Burke identified pleasure, along with pain, as being among the strongest of the emotions, though pain was by far stronger. Yet pleasure was connected to society and sex (as Erasmus Darwin concurred). Without moralizing, he wrote about the significance in humans of the use of reason in pursuing their beloved, while with other species lust was predominate in light of urgency during specific mating seasons. Burke argued that since humans have reason or a greater sense of sociability, they pause to seek out beauty; that is, men seek out beauty in women. Charles Darwin, of course, was interested in the concept of perceived beauty in women by men, though this would be developed only in his later work.[46] However, unlike Burke (and like Addison), Darwin would go on to propose an underlying sense of beauty in all creatures that in turn affects mate choice.

In discussing the passions, Burke was most taken with the sublime, and this would provide Darwin with important insights. Burke devoted Part II of his book to the sublime. Aspects of the sublime include among others astonishment, power, privation, infinity, terror, and obscurity (which we can understand literally as darkness or metaphorically, as a lack of comprehension of the workings, the shape of nature). Vastness, another characteristic, might have been of some interest to Darwin in his dilemma over biological clarity. Burke defined vastness as magnitude and brought up what was invisible to us in biology. He uses the microscope by way of example and wrote,

> When we attend to the infinite divisibility of matter, when we pursue animal life into these excessively small yet organized beings, ...when we push our discoveries yet downward…the still diminishing scale of existence, in tracing which the imagination is lost as well as the sense, we become amazed and confounded….[47]

These concepts of extension and infinity might offer parallels with the invisible and confounding nature of evolution.

Darwin puzzled over the physiology of fear, with its attendant aspects of pain, in the summer and fall of 1838; he became increasingly convinced that the expression of fear was based in ancestral histories of survival. Sensations of fear had become to some degree instinctive; they were accompanied by a beating heart and trembling muscles, which implied running away from the source that produced fear. One "starts" over a loud noise because the brain sends a mandate to the nerves and muscles. Disagreeable thoughts can result in the action of the brain creating the sensation of pain on muscles, for example, through twitching. Darwin was drawing implications that these physiological responses to external sources were based on actual actions to move away from or repel the source of fear in ancient populations. Twitching muscles, shaking (in the deep past literally shaking off a predator), and rapid breathing had now become hereditary responses as opposed to an actual course of action. The very expression of the passions was dependent on this history. He noted, "Expression is an hereditary habitual movement consequent on some action, which the progenitor did, when excited or disturbed by the same cause, which 'now' excites the expression."[48] When Darwin addressed the history of the passions as found in humans today in *The Expression of the Emotions*, fear was among the most ancient:

> …fear was expressed from an extremely remote period, in almost the same manner as it now is by man; namely by trembling, cold perspiration, pallor, widely opened eyes, the relaxation of most of the muscles and by the whole body cowering down or held motionless. Suffering, if great, will from the first have caused screams or groans to be uttered, the body to be contorted and the teeth to be ground together. But our progenitors will not have exhibited those highly expressive movements of the features which accompany screaming and crying until their circulatory and respiratory organs, and the muscles surrounding the eyes, had acquired their present structure.[49]

Burke did not venture in the direction of ancestral populations, but his focus on physiological response to the idea of terror and pain as central to survival and as largely inherent was in keeping with Darwin's general speculations on the struggle for life.

While the sublime was attached to preservation of the self, there was also society to consider and here Burke stressed pleasure. His two-fold approach addressed the positive and central passions associated with sex and reproduction or "generation" (which Erasmus Darwin applied to the very transformation of species), and the passions that led to "general society."

Charles Darwin was drawn to Burke's thoughts on our interest in the concerns of others and to his analysis of how aiding others in distress was often accompanied by delight in not being in dire circumstances oneself (which was, paradoxically, an aspect of the sublime).

Darwin's first marginalia in his personal copy of *Philosophical Enquiry* begins with Section XIII of Part I, "Sympathy," and here Burke writes of engagement in the emotions of others wherein if the emotion is distressing, it would be attached to the sublime or if pleasant, it would then be a source of "social affection." Through sympathy "we are put into the place of another man, and affected in many respects as he is affected." Burke then brings up the common misunderstanding that when we see or read of a tragic incident in painting or literature and feel sympathy, it is because we are supposedly relieved by its fiction, contemplating the distant evils therein, but that the real event would "shock." Darwin bracketed the following passage in Burke:

> I am afraid it is a practice much too common in inquiries of this nature, to attribute the cause of feelings which merely arise from the mechanical structures of our bodies, or from the natural frame and constitution of our minds, to certain conclusions of the reasoning faculty on the objects presented to us; for I should imagine, that the influence of reason in producing our passions is nothing near so extensive as it is commonly believed.[50]

Burke continues, "…I am convinced we have a degree of delight, and no small one, in the real misfortunes and pains of others."[51] Darwin's earliest notes in Notebook M from *A Philosophical Enquiry* (July 22, 1838) come from the next section, Section XIV of Part I, "The effects of sympathy in the distresses of others." Here Darwin contemplated the tendency to take pleasure in the distress of individuals who suffer. Burke believed that the "delight" experienced was a motivating factor that "[hindered] us from shunning scenes of misery."[52] In fact, this quality of delight, which Burke was at pains to distinguish from pleasure, was central to the sublime experience. According to Burke, "When danger or pain press too nearly, they are incapable of giving any delight, and are simply terrible; but at certain distances, and with certain modifications, they may be, and they are delightful, as we every day experience."[53] Crucially for Burke, sublime delight in situations of distress to others is a motivating factor: "…the pain we feel prompts us to relieve ourselves in relieving those who suffer." For Burke the reasoning behind the twining of delight and pain in this situation could be found in providence: "As our Creator has designed we should be united by the bond of sympathy, he has strengthened that bond by a proportionate delight; and there most where our sympathy is most wanted; in the distress of others."

Darwin recalled Burke on sympathy once again several weeks later on August 16: "[Remember] Burke's idea of Sympathy being real pleasure at pain of others, with rational desire to assist them—otherwise as he remarks sympathy could be barren & lead people from scenes of distress…"[54] Without citing providence, Darwin found this to be a far more logical explanation than that offered by Adam Smith, noting "Adam Smith…says 'sympathy' we can only know what others think by putting ourselves in their situation, & then we feel like them—, hence sympathy very unsatisfactory because it does not like Burke explain pleasure."[55] On August 16 Darwin also wrote extensively about shared passions among various species and brought up the orangutans at the Zoological Gardens, noting, "He who understands baboon will do more towards metaphysics than Locke." More than 30 years later, while exploring the concept of sympathy in *The Descent of Man* (1871), which he argued was of central importance to group survival, Darwin stressed its origins in animal society. In turning to the issue of coming to the aid of others in humans (though he had discussed this in animals as well), he brought up the speculations of physiological psychologist Alexander Bain: "Instinctive sympathy would also cause [man] to value highly the approbation of his fellow-men, for as Mr Bain has clearly shewn, the love of praise and the strong feeling of glory, and the still stronger horror of scorn and infamy are due to the workings of sympathy."[56] In turn, from Darwin's perspective, this would become a hereditary tendency.[57] The evolutionary trajectory might be as follows:

> …the social instincts, which must have been acquired by man in a very rude state, and probably even by his early ape-like progenitors, still give the impulse to many of his best actions; but his actions are largely determined by the expressed wishes and judgment of his fellow-men, and unfortunately still oftener by his own strong, selfish desires. But as the feelings of love and sympathy and the power of self-command become strengthened by habit, and as the power of reasoning becomes clearer so that man can appreciate the justice of the judgments of his fellow-men, he will feel himself impelled, independently of any pleasure or pain felt at the moment, to certain lines of conduct. He may then say, I am the supreme judge of my own conduct….[58]

Although Darwin does not mention Burke in this important section in *The Descent* on "moral sense," and "approbation," the philosopher may have helped provide the underlying premise to pursue sympathy from this perspective. Burke had written that the "three principal links in [the chain of society in general] are sympathy (discussed above), imitation (as in entering into a similar state of mind as others, also discussed above), and ambition." Ambition, according to Burke, was "a satisfaction arising from the contemplation of...excelling [over one's] fellows in something deemed valuable amongst them."[59] This could be interpreted as generous and singular acts of aid to others. As necessary as it was to society, ambition had a sublime side and could engage with the "selfish" passions concerned with self-preservation. Here Burke invokes that great early author on the sublime, Longinus, and the importance he credits to "glorying and sense of inward greatness."[60]

The relevance of considering Burke on ambition is reflected in Darwin's notations on the front and back covers of his copy of *Philosophical Enquiry*. On the front cover Darwin had written, "The morality & Metaphysics of ambition?" and "Sublimity?" On the back cover he wrote, "Simple Ambition instinct of excellence over other men satisfied (1)." The attendant emotions of approbation are outlined: Pride, Fame, Vanity, Arrogance, and Conceit. He continues, "The feeling of Sublimity akin to feeling of pure (1) gratified ambition…"[61] In this case, Darwin has connected the sublime to the individual and to grand acts that effect the self (e.g. through praise and admiration), though its roots from Burke's perspective and ultimately from his own were in society.

At the end of Part II on the sublime in *Philosophical Enquiry*, Burke summarized the biological sublime, concluding, "Having thus run through the causes of the sublime with reference to all the senses, my first observation... will be found to be nearly true; that the sublime is an idea belonging to self-preservation. That it is therefore one of the most affecting we have."[62] Burke's *Enquiry* provided speculations on the underlying nature and expression of the passions in regard to individual survival and "society" that lack the socially disapproving tone of Malthus or certain influential moral philosophers during a critical juncture in Darwin's development. Such passions were not entirely antithetical to those of animals as observed by Stubbs. In addition, Burke presented an argument in keeping with the unity of mankind and underlying mental faculties without relying on the associationism so popular among Erasmus Darwin's generation. Sympathy, for example, arises from "the mechanical structure of our bodies, or the natural constitution of our minds." And while Burke did not venture into ancestral discussions, his ideas on the physiology of pain, delight, and pleasure and its personal and social implications provided observations that were close to the interests of the Charles Darwin.

Acknowledgements

I would like to thank Janet Browne, Evelleen Richards, and Nicholas Power for their helpful comments on this chapter and Nicola Gauld for her research assistance.

Notes

1. James Paradis, "Darwin and Landscape," in *Victorian Science and Victorian Values: Literary Perspectives*, eds James Paradis and Thomas Postlewait (New Brunswick, NJ: Rutgers University Press, 1985), 85–110; David Kohn, "The Aesthetic Construction of Darwin's Theory," in *The Elusive Synthesis: Aesthetics and Science*, ed. Alfred Tauber (The Hague: Kluwer, 1996), 13–48; Phillip Sloan, "'The Sense of Sublimity': Darwin on Nature and Divinity," *Osiris*, 2nd series, 16, 251–69; and Robert Richards, *The Romantic Conception of Life: Science and Philosophy in the Age of Goethe* (Chicago: University of Chicago Press, 2002).

2. Charles Darwin, *The Autobiography of Charles Darwin, 1809–1882*, ed. Nora Barlow (New York: The Norton Library, 1969), 85.

3. Alexander von Humboldt and Aimé Bonpland, *Personal Narrative of Travels to the Equinoctial Regions of the New Continent During the Years 1799–1804*, trans.. H.M. Williams, 7 vols (London: Longman Hurst, 1818–1829). Although there are two authors, Darwin exclusively referred to Humboldt's evocative descriptions. On the specific volumes, see "Books on Board the *Beagle*," appendix IV, in *The Correspondence of Charles Darwin*, eds Frederick Burkhardt, Sydney Smith, David Kohn, and William Montgomery (Cambridge: Cambridge University Press, 1985), 1, 561.

4. Darwin's annotated copy of Burke's *Enquiry* is in the rare books collection of the Cambridge University Library (Adv.d.79.21) and, as of this date, not housed with the collection at the Darwin Library there. I thank Evelleen Richards for bringing it to my attention. Published by Thomas McLean in 1823, the edition was presented to Darwin in 1828 and inscribed: "G.V. Jackson to Charles Darwin Christ Coll April 1828." References to the Burke appear in Notebooks M and N, which investigate the general topic of the origins of behavior. Notebook M contains notes from July 15, 1838 to October 2, 1838, many of which refer to conversations with his father Robert on human thought and behavior. Robert was a medical practitioner, as had been the case with Erasmus Darwin and both the elder Darwins tied medicine to broader philosophical and behavioral speculations. Notebook N dates largely from October 2, 1838 to July 20, 1839, with a few later additions (up to April 28, 1840). The notebook entries are published in *Charles Darwin's Notebooks, 1836–1844: Geology, Transmutation of Species, Metaphysical Enquiries*, transcribed and ed. Paul Barrett, Peter Gautrey, Sandra Herbert, David Kohn, and Sydney Smith, (Ithaca, NY: Cornell University Press, 1987).

5. Addison's writings in *The Spectator* in 1712, together referred to as "Pleasures of the Imagination," not only made the sublime (though the specific term was not used) a central consideration of aesthetics, but also initiated the tradition of identifying the reading of *Paradise Lost* as the greatest literary source to stimulate the imagination. Specifically on Milton is paper 417, published in *The Spectator* on June 28, 1712.

6. *The Correspondence of Charles Darwin*, 1, 237. The two works he referred to were Johann Moritz Rugendas' lithograph, *Brazilian Forest*, 1828, and Claude-Francois Fortier's engraving of 1822 after Charles-Othon-Frédéric-Jean-Baptiste comte de Clarac's drawing, *Brazilian Forest*, of 1819.

7. R.D. Keynes, ed., *Charles Darwin's* Beagle *Diary* (Cambridge: Cambridge University Press, 2001), 114.

8. Darwin, *Autobiography*, 90.

9. Ibid., 91.

10. Ibid., 30.

11. Francis Darwin, ed., *The Life and Letters of Charles Darwin, Including an Autobiographical Chapter*, vol. 3 (London: John Murray, 1887), 67.

12. Edmund Burke, *A Philosophical Enquiry into the Origin of Our Ideas of the Sublime and Beautiful*, ed. Adam Phillips (Oxford: Oxford University Press, 1990), 1.

13. On sensibility in the eighteenth century, see Janet Todd, *Sensibility: An Introduction* (New York: Methuen, 1986); F. Hilles and H. Bloom, eds, *From Sensibility to Romanticism: Essays Presented to Frederick A. Pottle* (New York: Oxford University Press, 1965); Louis Bredvold, *The Natural History of Sensibility* (Detroit: Wayne State University Press, 1962), and John Mullan, *Sentiment and Sociability: The Language of Feeling in the Eighteenth Century* (New York: Clarendon Press, 1988); G.J. Barker-Benfield, *The Culture of Sensibility: Sex and Society in Eighteenth-Century Britain* (Chicago: University of Chicago Press, 1992).

14. On Nugent and Burke, see Aris Sarafianos, "The Contractility of Burke's Sublime and Heterodoxies in Medicine and Art," in *Journal of the History of Ideas* 69, 1 (January 2008), 23–48 and her "Pain, Labour and the Sublime: Medical Gymnastics and Burke's Aesthetics," *Representations* 91 (Summer 2005), 58–83. Sarafinos has also pointed out the influence of Burke's schoolmate and friend Richard Brocklesby, whose experiments on animals in pain led to his conclusion that skin and nerves were sensitive but immobile while muscle fibers and membranes were insensible but motile. She argues that Burke extrapolated from this physiological division of pain and insensibility *aesthetic* polarities between pain and pleasure. Pleasure was associated with the beautiful and pain with the sublime ("Pain, Labour, and the Sublime," 63).

15. See Vanessa Ryan, "The Physiological Sublime: Burke's Critique of Reason," *Journal of the History of Ideas* 62, 2 (April 2001), 276.

16. Burke, *A Philosophical Enquiry*, 59–60. The principles of pleasure and pain had long been considered central to the passions and aesthetics, developed in the writings of Locke. Locke had

posited that they were necessarily intertwined, pleasure being produced by a lessening of pain and vice versa, but Burke's position is that they were often independent. In *A Philosophical Enquiry*, he maintained that a lessening of pleasure often leads to indifference and a lessening of pain to "consternation" or "sobriety."

17. Ibid., 119.

18. Burke was by no means the only theorist of physiological response that Darwin consulted on the "passions" in these early years. Other sources of the late 1830s include Charles Le Brun's *Conférence sur l'expression générale et particulière* (1698) and Charles Bell's publications. However, the thesis of the driving force of self-preservation is central only to Burke.

19. On the commission, see Malcolm Warner and Robin Blake, *Stubbs and the Horse* (New Haven, CT: Yale University Press, 2004), 109. By December of 1765, Burke, a liberal Whig, became a member of British Parliament. On Burke and Stubbs, also see Aris Sarafianos, "Stubbs, Walpole, and Burke: Convulsive Imitation and 'Truth Extorted'," *Tate Papers* 13 (Spring 2010). Accessed May 2011. http://www.tate.org.uk/research/tateresearch/tatepapers/10spring/sarafianos.shtm.

20. Burke, *Enquiry*, 60–61.

21. There has been some controversy over the identity of the sitter, but the Wedgwood Museum, where the portrait resides, continues to identify it as Erasmus Darwin.

22. Desmond King-Hele, *Erasmus Darwin: A Life of Unequalled Achievement* (London: Giles de la Mare Publishers, 1999), 353.

23. George Stubbs, *A Comparative Anatomical Exposition of the Human Body, with that of a Tiger and a Common Fowl*. The 126 drawings are catalogued in *The Anatomical Works of George Stubbs* by Terrence Doherty (London: Martin Secker & Warburg Ltd, 1974).

24. Fuseli was an amateur entomologist and frequented medical circles from the 1760s on, where he may have met Erasmus Darwin. *The Nightmare*, though it appears macabre, was partly inspired by current medical theories; nightmares in this context were also of interest to Erasmus Darwin.

25. An engraved version of Fuseli's *Nightmare* appears in certain editions of *The Loves of the Plants* (1789), which is the first part of *Botanic Garden*; the *Economy of Vegetation* being the second part.

26. Burke, *Enquiry*, 54.

27. Rousseau had originally written about this in his *Discourse on Equality* (1755) and elsewhere. Christopher Frayling goes so far as to suggest that the *mara* figure may be a reference to crossbreeding, one aspect of potential relatedness that Rousseau had suggested. See his essay, "Fuseli's *The Nightmare*: Somewhere between the Sublime and the Ridiculous," in *Gothic Nightmares: Fuseli, Blake and the Romantic Imagination*, Martin Myrone, ed. (London: Tate Publishing, 2006), 9–20.

28. On self-protection and the passions, see, for example, *The Temple of Nature, or, the Origin of Society: A Poem with Philosophical Notes* (1803) (Menston, Yorkshire: Scolar Press), facsimile of the original (1973), 113.

29. Erasmus Darwin, *Zoonomia, or the Laws of Organic Life*, 2 vols (London: J. Johnson, 1794), 3rd edn, 4 vols. The citation is from the 3rd edn, 2, 12: I, 67, 73.

30. On Erasmus Darwin, sex, generation, and species change see Roy Porter, "Erasmus Darwin: Doctor of Evolution?," in *History, Humanity, and Evolution: Essays for John C. Greene*, ed. James R. Moore (Cambridge: Cambridge University Press, 1989), 39–70.

31. Burke, *Enquiry*, 105.

32. Erasmus Darwin, *The Temple of Nature*, "Additional Notes," 91.

33. Ibid., 91.

34. Erasmus Darwin, *Zoonomia*, I, 507.

35. On the problematic dating of this notebook, see David Kohn, "Notebook B," in *Charles Darwin's Notebooks*, 167–9.

36. Darwin, *Autobiography*, 120.

37. This is written in Notebook C. See *Notebooks*, 264.

38. *Zoonomia*, I, XVI (Of Instinct): VII.

39. *Notebooks*, 566. On October 4, 1838, Darwin had recorded this comment after reading Lavater (*Essays on Physiognomy*, 2nd edn, 1804, III: 37–38), but he may well have been reminded of his grandfather's interest in it.

40. On the development of conscience in Darwin's thinking and his reading in the late 1830s, see Robert Richards, *Darwin and the Emergence of Evolutionary Theories of Mind and Behavior* (Chicago: University of Chicago Press, 1987), especially 185–242.

41. Burke, *Enquiry*, 118.

42. Burke's interest in the innate is related to the Scottish school of mind, which posited that innate mental faculties like curiosity, attention, and perception exist prior to sensory input. See James McCush, *The Scottish Philosophy: Biographical, Expository, Critical, from Hutcheson to Hamilton* (New York: AMS Press, 1980), and Louis Schneider, ed., *The Scottish Moralists on Human Nature and Society* (Chicago: Chicago University Press, 1967).

43. On Darwin's early use of associationism conjoined with habitual behavior to form a theory of inherited unconscious instinct, see Robert Richards, "Influence of Sensationalist Tradition on Early Theories of the Evolution of Behavior," *Journal of the History of Ideas* 40, 1 (January–March 1979), 85–105. Later this would be modified to accommodate natural selection, wherein instincts like structures are subject to selection.

44. *Notebooks*, 531.

45. Ibid., 567.

46. Burke refuted Addison on the importance of the concept of beauty in mating among non-human species, but it would be accepted and central to Darwin's theory of sexual selection.

47. Burke, *Enquiry*, 66.

48. *Notebooks*, 545.

49. Charles Darwin, *The Expression of the Emotions in Man and Animals*, ed. Paul Ekman (Oxford: Oxford University Press, 1998), 356.

50. Burke, *Enquiry*, 41.

51. Burke, *Enquiry*, 42.

52. Burke, *Enquiry*, 43.

53. Burke, *Enquiry*, 36–7.

54. *Notebooks*, 540.

55. *Notebooks*, 546.

56. Charles Darwin, *The Descent of Man, and Selection in Relation to Sex* (1871) (Princeton, NJ: Princeton University Press, 1986), 86.

57. Ibid., 161–7.

58. Ibid., 86.

59. Burke, *Enquiry*, 46. The concept of approbation had also been mentioned by Smith in his *The Theory of Moral Sentiments* (1759). Burke felt that such a theory was in keeping with his own program. His review of Smith's book in *The Annual Register* is discussed in Daniel O'Neill, *The Burke–Wollstonecraft Debate: Savagery, Civilization, and Democracy* (College Park, PA: Penn State University Press, 2007), 63–6.

60. Burke, *Enquiry*, 46.

61. For full annotations, see Mario di Gregorio with the assistance of N.W. Gill, *Charles Darwin's Marginalia* 1 (New York: Garland, 1990) 102–103.

62. Burke, *Enquiry*, 79.

2

Why is the peacock's tail so beautiful?

Laurence Shafe

In the second half of the nineteenth century, beauty was an important issue. Fyodor Dostoevsky (1821–1881), loosely associated with the Aesthetic Movement, wrote in his notebook, "Beauty will save the world—two kinds of beauty."[1] But what were these two kinds of beauty? A hint is provided in his final novel of 1880, *The Brothers Karamazov*. One of the brothers, Dmitry, reads a poem which includes the line "And lust in lowly insect fires" and he adds, "I am that insect" and goes on to say, "Beauty! It makes me mad to think that a man of great heart and high intelligence should begin with the ideal of Madonna and end with the ideal of Sodom."[2] This could be a metaphor for the journey from John Ruskin's (1819–1900) view of divine beauty through Charles Darwin's (1809–1882) theories of beauty and animal sexuality to the accusation of sodomy brought against the aesthete and devotee of beauty, Oscar Wilde (1854–1900) in 1895, whose subsequent trial and incarceration has traditionally marked the terminal point of the Aesthetic Movement. While beauty was interpreted in various ways in the final decades of the nineteenth century, the peacock became a reference point for discussions on creation, evolution, and the aesthete's taste for decadent refinement without allusion to divinity or scientific rationale. Its appeal as the ultimate aesthetic symbol to all camps is the subject of this chapter.

Darwin's *On the Origin of Species* was published at the end of 1859 and quickly became immensely popular. During the next ten years, two animals came to symbolize his ideas in the popular imagination—the ape and the peacock. The ape is perhaps the better-known symbol, as it brought our animal origins, with their ludicrous and salacious associations, forcibly to the public's attention. The peacock, on the other hand, had long been a symbol of beauty and Darwin's claim that the spectacular tail was used to attract mates was controversial and opposed by many, including Ruskin and George Campbell, 8th Duke of Argyll (1823–1900).[3] Both regarded beauty, such as that of the peacock's tail, as proof of divine creation and they thought Darwin's theory offered no acceptable explanation of how it could have come about.[4]

Even Darwin found the peacock's tail a problem. He wrote to Asa Gray in 1860, "The sight of a feather in a peacock's tail, whenever I gaze at it, makes me sick!,"[5] and in 1864, he wrote to Alfred Wallace, "It is an awful stretcher to believe that a Peacock's tail was thus formed," but significantly he added, "but believing it, I believe in the same principle somewhat modified applied to man."[6] Darwin's explanation for how beauty evolved was sexual selection, but it was not until 1871, when he published *The Descent of Man and Selection in Relation to Sex* that he explained his ideas fully. The book was essentially two books in one, as the title indicated, and much of the book was devoted to sexual selection. Sexual selection is concerned with two types of physical characteristics: *weapons* that one gender uses to dominate other members of the same gender and drive off rivals and *ornaments* that one gender uses to attract the other. Both weapons and ornaments are concerned with maximizing mating opportunities. There are also many evolutionary stable strategies employed by species as part of the mating game, such as lifetime pairing, lekking and polygyny.[7] Natural selection, concerned with maximizing food sources and avoiding enemies, and sexual selection often pull in opposite directions as we see with the peacock's tail, which requires extra food to produce and makes the owner less able to escape predators. Poetically, this can be seen as a battle between avoiding death and cultivating beauty.

In *Descent of Man*, Darwin grappled with the difficult problem of beauty in humans. In all animals in which the female devotes considerable resources to the nurturing of the young, it is the male that possesses the secondary sexual characteristics involving display and either fights for possession of the female using weapons, or attracts the female using beauty. Darwin pointed out that in humans, both men and women possess secondary sexual characteristics, implying that both sexes were involved in selection, though ultimately men gained an advantage in choice.

The subject of sexually assertive females was explored by artists of the Aesthetic Movement, such as Dante Gabriel Rossetti (1828–1882) and Frederic Leighton (1830–1896) during the 1860s. Rossetti's first "female heads with floral attributes" was *Bocca Baciata,* painted during September and October 1859, in the period between Darwin's paper on the transmutation of species being read to the Linnaean Society (July 1, 1858) and the publication of *Origin* (November 25, 1859).[8] *Bocca Baciata* means "kissed mouth" and refers to Novel 7 on the Second Day of *Decameron* by Giovanni Boccaccio (1313–1375), in which a sexually assertive woman travels through a remote country seducing eight men until she finally returns to marry the king as a virgin.[9] Also during this period in 1859, Leighton was in Rome and produced a number of paintings of Nanna Risi, a popular model whose large, imposing, statuesque figure and commanding looks were portrayed by many artists. Leighton painted her

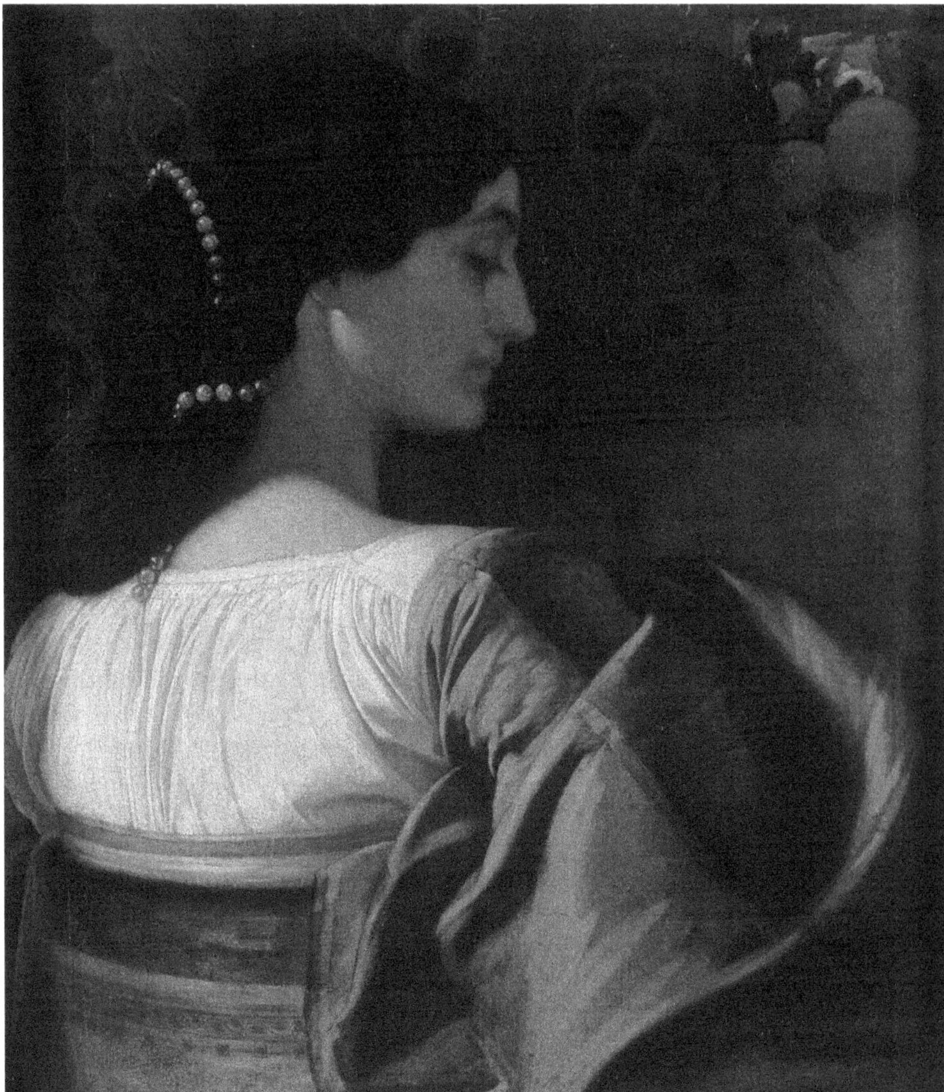

2.1 Frederic Leighton, *Nanna (Pavonia)* (1859) oil on canvas, 59.4 × 51.1 cm. The Royal Collection, © 2011 Her Majesty Queen Elizabeth II

over the winter of 1858–1859 and his art works were admired by the Prince of Wales during a visit to his studio. Three of the paintings were displayed in 1859 at the Royal Academy. One, a small painting called *Summer Hours*, is now lost, another is in the Royal Collection, and the third is in the Philadelphia Museum of Art. The painting in the Royal Collection, *Nanna (Pavonia)* (Figure 2.1), had already been sold to Georges de Monbrison, but Leighton persuaded him to give it to the Prince of Wales—later Edward VII—in exchange for painting a copy.[10] The copy, called *Pavonia* (Italian for peahen), is now at Leighton House and both paintings show a female figure holding a peacock feather fan.

By incorporating the Latin word for peacock, *pavo*, into the title of the painting, Leighton associated the woman with the beauty of the bird.[11] The halo of peacock feathers is part of a fixed fan that would have been seen as exotic, as it was imported from India, and it had both literary and classical associations.[12] Fans were a common fashion accessory that from the sixteenth century onwards were used by men and women for flirting and as a means of communication referred to as "fanology."[13] In the nineteenth century, fans continued to be an important social tool and their use included hiding the face to demonstrate coyness. In *Nanna (Pavonia)* a dark-eyed Mediterranean woman turns in her chair not to look at us, but as if to look away from someone else in the room, possibly using the fan to hide her face and therefore giving us a rarely seen glimpse of a coy woman at her most secretive moment.

Darwin wrote that the female is sexually "less eager than the male" and quoted John Hunter (1728–1793) as saying that "she generally 'requires to be courted;' she is coy."[14] Darwin reinforced this idea of coyness with a story. Imagine, he wrote, the "inhabitants of another planet" watching "young rustics at a fair, courting and quarrelling over a pretty girl, like birds at one of their places of assemblage" and the aliens concluding that "she had the power of choice."[15] Darwin was using human behavior to illustrate how a female bird selects a particular male at a lek, a location where peafowl gather for the females to choose the most beautiful males to mate with. He introduces the alien as an impartial observer, suggesting the conventional view at the time would be that the girl was being selected by the young rustics and had no choice of her own. Darwin believed that in spite of ultimate male dominance in this area, "in civilized nations women have free or almost free choice" between those attracted to her, and among what he called "savages," the women "can tempt the men who they prefer, and can sometimes reject those whom they dislike."[16] The idea that women had any real choice was not widely taken for granted as it was assumed, even by Darwin, that men were more rational than women. This meant moral order could be undermined because it was assumed it was based on restraint and rational control. However, coyness as a social strategy implies sexual attraction and seduction in the guise of modesty and so Darwin's story supported the idea of the strong, manipulative woman at the same time as artists were portraying their version—the *femme fatale*.

Many artists were exploring female sexuality during mid-century, but often with a moral lesson. For example, Frederick Sandys' *Vivien* (1863, Manchester City Art Gallery) shows the legendary enchantress of Merlin in Arthurian legend, modeled by Keomi Gray. She is also shown surrounded by a halo of peacock feathers, but this symbol of beauty is subverted into one of seduction by the use of other images including an apple, suggesting temptation, and a sprig of poisonous daphne. Social order during the Victorian period was thought to be maintained by male self-control and so the picture can be read as a moral tale—female beauty and enchantment are being used to achieve a level of sexual arousal that will overcome male moral restraint and lead to social, if not physical, death. Sandys' painting is, therefore, unlike Leighton's or Rossetti's, making a moral point, and we shall see that this role for art was rejected entirely by the Aesthetic Movement.

John Ruskin's reaction to Darwinian aesthetics

Darwin's evolutionary explanation for beauty based on sexual selection was a challenge to Ruskin's *theoretic faculty*, his term for the aesthetic—a word he objected to as he felt it overemphasized the sensual, given its Greek etymology. Ruskin completely rejected Darwin's ideas linking the aesthetic sense in humans with the physical sensations of animals, as he felt this provided a naturalistic and scientific basis for the promotion of materialism. However, Ruskin did not argue point for point against Darwin's views, but attempted to ridicule them, possibly because he knew he had little credibility as a scientist. For example, he wrote that, "the scientific world professed itself to have discovered that the mollusc was the Father of Man;" and while disingenuously claiming in a footnote that his ideas "are in nowise antagonistic to the theories which Mr Darwin's unwearied and unerring investigations are every day rendering more probable" he continued by asserting that, "the characteristic form of each species is entirely individual," which for Ruskin meant immutable.[17] In *The Eagle's Nest* he wrote, "Had Darwinism been true, we should have long ago [...] changed ourselves into Briarean Cephalopoda."[18] In *Mornings in Florence* he went as far as to say "Darwinism, like all popular and widely mischievous fallacies, has many a curious gleam and grain of truth in its tissue."[19] His attempts to make fun of Darwinism culminated in the widely quoted skit in *Love's Meinie* (1873), in which he suggested that by "fastening a hair-brush to a mill-wheel" it would, on hearing a steam-whistle, fall in love with the whistle, lay an egg, and produce a nightingale.[20] *Love's Meinie* was based on the Oxford lectures he gave two years after *Descent* was published, and in it he studied a number of birds and their feathers in order to describe their aesthetic and mythological richness compared with the "arrogance and materialism" of science.

Ruskin argued strongly against materialism; for example, by pointing out that he believed the imperfections and idiosyncrasies of the world prevented it being explained mechanically. He was nevertheless impressed by the perfection of machinery. In point seven of *The Story of Arachne* he described the "amazed awe" with which he watched a locomotive "take its breath at a railway station."[21] He continued by describing the many ways in which animals are imperfect, including the "ignoble" eating habits of eagles and the ineffectual "slobbering" of the pelican trying to eat. He contrasted this with his "much respected friend Professor Huxley," who asserted sight was a "mechanical operation."[22] Ruskin was associating the materialism of Darwin and Thomas Huxley (1825–1895) with a mechanical explanation of the natural world which, he argued, was too simplistic, as the world was messy and inexplicable. Huxley used the mechanical metaphor to suggest that component parts can be disassembled and their function understood without any mysterious or metaphysical explanation; for Ruskin, the metaphor implied a perfection of function that ran counter to his experience of the "awkwardness" of nature. By stressing the fallibility of animals, Ruskin was able to present them as impossible to explain mechanistically and by finding fault with Huxley's metaphor, he was able to imply that organisms ultimately were outside the scope of science and required a divine explanation. He referred to the same Huxley metaphor in *The Eagle's Nest*, although he only referred there to a "great physiologist" and he specifically added that, "sight is an absolutely spiritual phenomenon."[23] For Ruskin, the danger was a scientific materialism that sought to explain everything including the aesthetic, leaving no room for the spiritual and the moral.

Ruskin in his youth, like Darwin, had been drawn to geology and had attended Darwin's debut lecture in 1837 at the Geological Society, when he talked about the coast of Chile. Darwin was so pleased by his reception that he described himself as "like a peacock admiring his tail."[24] They met a few months later at William Buckland's (1784–1856) house in Oxford "and talked all the evening."[25] In 1879, they met for a second time in Ruskin's home at Coniston and Darwin admitted in private that he could make nothing of what Ruskin saw in his Turners, and Ruskin, who thought Darwin ignorant of good art and knew of his inability to draw, used this to undermine his arguments.[26] In *Deucalion* (1875–1883), Ruskin attacked Darwin, Huxley and John Tyndall (1820–1893) and went so far as to defend the value of myth as being "'incomparably *truer*' than the Darwinian—or, I will add, any other conceivable materialistic theory."[27] However, by the 1880s, Ruskin conceded that natural selection might be possible, although he never accepted it as a full explanation.[28]

The aesthetic movement and Darwin's aesthetics

One month after Darwin published *The Descent of Man,* Edward Linley Sambourne (1844–1910) satirized his ideas in *Punch* (Figure 2.2). The cartoon was part of a long-running series lampooning the grotesque effects that can be achieved by combining nature and fashion. The cartoon can be seen as an intellectual pivot point between the biological discussions of the 1860s and the

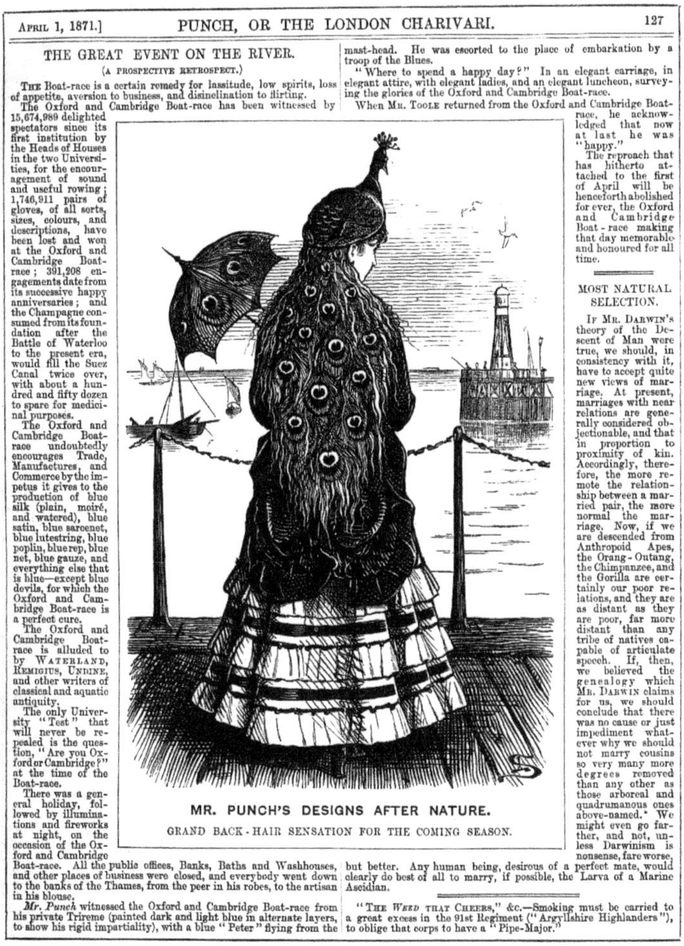

April 1, 1871.] PUNCH, OR THE LONDON CHARIVARI. 127

THE GREAT EVENT ON THE RIVER.

(A PROSPECTIVE RETROSPECT.)

The Boat-race is a certain remedy for lassitude, low spirits, loss of appetite, aversion to business, and disinclination to flirting.

The Oxford and Cambridge Boat-race has been witnessed by 15,674,989 delighted spectators since its first institution by the Heads of Houses in the two Universities, for the encouragement of sound and useful rowing; 1,746,911 pairs of gloves, of all sorts, sizes, colours, and descriptions, have been lost and won at the Oxford and Cambridge Boat-race; 391,208 engagements date from its successive happy anniversaries; and the Champagne consumed from its foundation after the Battle of Waterloo to the present era, would fill the Suez Canal twice over, with about a hundred and fifty dozen to spare for medicinal purposes.

The Oxford and Cambridge Boat-race undoubtedly encourages Trade, Manufactures, and Commerce by the impetus it gives to the production of blue silk (plain, moiré, and watered), blue satin, blue sarcenet, blue lutestring, blue poplin, blue rep, blue net, blue gauze, and everything else that is blue—except blue devils, for which the Oxford and Cambridge Boat-race is a perfect cure.

The Oxford and Cambridge Boat-race is alluded to by Waterland, Remigius, Undine, and other writers of classical and aquatic antiquity.

The only University "Test" that will never be repealed is the question, "Are you Oxford or Cambridge?" at the time of the Boat-race.

There was a general holiday, followed by illuminations and fireworks at night, on the occasion of the Oxford and Cambridge Boat-race. All the public offices, Banks, Baths and Washhouses, and other places of business were closed, and everybody went down to the banks of the Thames, from the peer in his robes, to the artisan in his blouse.

Mr. Punch witnessed the Oxford and Cambridge Boat-race from his private Trireme (painted dark and light blue in alternate layers, to show his rigid impartiality), with a blue "Peter" flying from the mast-head. He was escorted to the place of embarkation by a troop of the Blues.

"Where to spend a happy day?" In an elegant carriage, in elegant attire, with elegant ladies, and an elegant luncheon, surveying the glories of the Oxford and Cambridge Boat-race.

When Mr. Toole returned from the Oxford and Cambridge Boat-race, he acknowledged that now at last he was "happy."

The reproach that has hitherto attached to the first of April will be henceforth abolished for ever, the Oxford and Cambridge Boat-race making that day memorable and honoured for all time.

MR. PUNCH'S DESIGNS AFTER NATURE.

GRAND BACK-HAIR SENSATION FOR THE COMING SEASON.

MOST NATURAL SELECTION.

If Mr. Darwin's theory of the Descent of Man were true, we should, in consistency with it, have to accept quite new views of marriage. At present, marriages with near relations are generally considered objectionable, and that in proportion to proximity of kin. Accordingly, therefore, the more remote the relationship between a married pair, the more normal the marriage. Now, if we are descended from Anthropoid Apes, the Orang-Outang, the Chimpanzee, and the Gorilla are certainly our poor relations, and they are as distant as they are poor, far more distant than any tribe of natives capable of articulate speech. If, then, we believed the genealogy which Mr. Darwin claims for us, we should conclude that there was no cause or just impediment whatever why we should not marry cousins so very many more degrees removed than any other as those arboreal and quadrumanous ones above-named.* We might even go farther, and not, unless Darwinism is nonsense, fare worse, but better. Any human being, desirous of a perfect mate, would clearly do best of all to marry, if possible, the Larva of a Marine Ascidian.

"The *Weed* that Cheers," &c.—Smoking must be carried to a great excess in the 91st Regiment ("Argyllshire Highlanders"), to oblige that corps to have a "Pipe-Major."

2.2 "Mr Punch's Designs after Nature: Grand Back-Hair Sensation for the Coming Season," *Punch, or the London Charivari*, 1 April 1871, 127

aesthetic arguments that followed. We see the back of a woman parading at the seaside with a peacock parasol and hat, the feathers of which are enmeshed with her long flowing hair. The woman's sexual display is enhanced by her adoption of a male peacock's sexual characteristic, perhaps signifying her recognition that she is being chosen for her beauty, but the significance is ambiguous, as her gaze controls a symbol of male potency—the lighthouse. Sambourne satirizes the pretensions of Darwin's supporters, the affectation of ideas about beauty, the aspirations of the women's emancipation movement, and the grotesqueness of fashion that at this period often involved whole birds nestled on fans and posed on hats.

There is also a more subtle point made through the association of Darwin's ideas with fashion. Darwinian sexual selection is not necessarily based on utility, but often on idiosyncratic preference for certain undirected variations, although it took another hundred years for Russell Lande (b. 1951) to show mathematically that male trait and female preference can lead to runaway evolution based on a chance variation at the start.[29] There is, therefore, no explanation for particular forms of beauty; they can be completely random, like fashion.[30]

Darwin's aesthetic theory, or theory of beauty, was argued against constantly and a serious opponent was St George Mivart (1827–1900), who published *On the Genesis of Species* the same year as *Descent*. It was a well-argued rebutal in which he wrote,

> Mr Darwin explains the brilliant plumage of the peacock or the humming-bird by the action of sexual selection: the more and more brilliant males being selected by the females (which are thus attracted) to become the fathers of the next generation, to which generation they tend to communicate their own bright nuptial vesture. But there are peculiarities of color and of form which it is exceedingly difficult to account for by any such action.[31]

One example he quotes is Darwin's five geographically separated reports of the sudden appearance of a black-shouldered peacock as evidence against his own theory, as it required small changes taking place over long periods of time, not the sudden birth of an entirely new species.[32] Darwin was one of the first scientists to use rigorously what today is regarded as the scientific method, and this included presenting all the arguments and evidence against his own theory and then dealing with each in turn. In this case, he wrote, "I have heard of no other such cases in the animal or vegetable kingdom," which implies the observation is possibly inaccurate and he went on to suggest it was not a new species but a variety and if it occurred, the black-shouldered form could be a reversion to a previous type. However, he considered this as unlikely as finding a domestic dog that had only mated with other dogs suddenly giving birth to a fully formed wolf.

Although we may regard these scientific arguments as esoteric, there were many magazines that featured such scientific discussions alongside articles on art, philosophy, and poetry. They were read by a large cross-section of middle-class society. Examples include the *Athenaeum, Blackwood's Magazine, Chambers's Journal, Punch,* and *Macmillan's Magazine,* and together just these magazines had a circulation of about 125,000. The articles were not popularized versions of essays published elsewhere, as they often are today, but serious, in-depth primary sources of critical issues and this led to the cross-cultural spillover of themes from one discipline to another.

There were also light-hearted references to the new aesthetics; for example, as early as 1867, *Punch* had an article decrying the slaughter of birds such

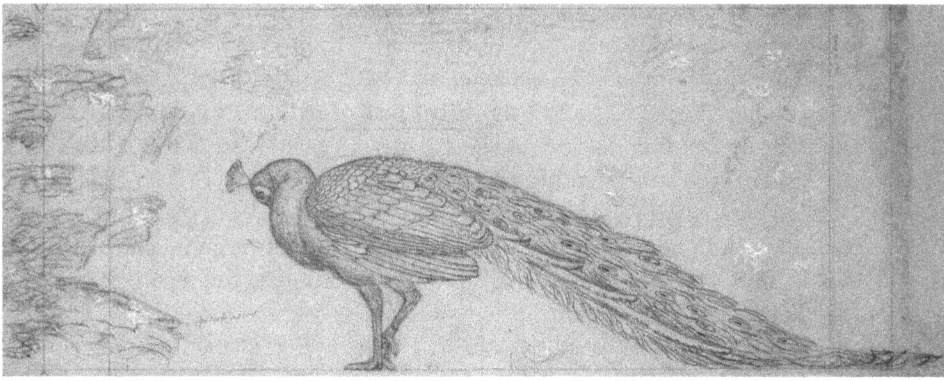

2.3 Albert Moore, cartoon for a portion of the Peacock Frieze executed in 1872 for the house of A.F. Lehmann, 15 Berkeley Square, charcoal and white chalk on brown paper, 56 × 129 cm, © Victoria and Albert Museum, London

as the peacock and birds of paradise so that their feathers could be used for "mantles, muffs and bonnets."[33] The 1860s and 1870s were decades in which peacocks were popular both within fine art, such as Leighton's *A Girl Feeding Peacocks* (c. 1863), as well as in the decorative arts, including homes that were decorated with peacock feathers, peacock wallpaper and soft furnishings, such as William Morris's *Peacock and Dragon* design (1878, The Textiles Collection: University for the Creative Arts at Farnham) and Arthur Silver's (1853–1896) *Peacock Feather* fabric for Liberty's (*c.* 1887).

Two major peacock-theme interiors were Albert Moore's design for Augustus Frederick Lehmann's house at 15 Berkeley Square in 1872 (Figure 2.3), and James McNeill Whistler's (1834–1903) better known and controversial Peacock Room, an interior he designed for Leyland House at 49, Princes Gate in 1876–1877. Moore worked with the architect and interior designer George Aitchison to design an asymmetrical frieze of peacocks in various natural poses, possibly sketched from nature. Moore was a methodical craftsman, who typically used multiple sketches that he worked up to life-size cartoons, and he had already used the peacock motif in 1861 for the panels of a fountain he designed for Croxteth (the home of the Molyneux family) and in his painting, *A Girl Dancing* (1863–1864).[34] Moore's peacock frieze was one of the earliest uses in interior design of a symbol that came to fill many an Aesthete's home over the next decade. His peacocks are unique in combining naturalism within a constrained geometrical format.

The most influential supporter of Darwin's theory of beauty was Grant Allen (1848–1899), science writer, novelist and an early pioneer of science fiction. Ruskin wrote, "Why we receive pleasure from some forms and colours, and not from others, is no more to be asked or answered than why we like sugar and dislike wormwood," but in *Physiological Aesthetics* (1877),

Allen challenged Ruskin's assumption by investigating the basis of our experience of pleasure and pain, and from this he built a theory of aesthetics.[35] By taking this approach he believed he could remove what he regarded as transcendental rhetoric and vague poetical declamations and arrive at a scientific theory that could be tested and which showed that our likes and dislikes are the result of natural selection.

Allen started by explaining that pain has evolved to force us to remove ourselves from sources that could damage valuable organs. He provided, like Ruskin, an extended metaphor of a locomotive but went further by speculating that, if conscious, it would experience pain at the loss of a piston or discomfort if short of coal or water. This equating of the mechanical with the animal was the link that Ruskin refused to accept. Allen related pleasure to the healthy functioning of the organs, as long as this was not taken to excess, and he believed it was the business of art to combine as many as possible of the pleasurable sensations and exclude the painful to produce what he called the "aesthetic thrill." His definition of the aesthetically beautiful was that which afforded "the Maximum of Stimulation with the Minimum of Fatigue or Waste" and the aesthetically ugly was that which failed to do so.[36]

Allen considered that we derive more pleasure from the nude figure, as the graceful natural outline was not masked by clothing, but he found tinted statues, despite the Hellenic precedent, offended the "average modern taste," as they threatened the spectator with the "obtrusion of non-aesthetic sentiments." He pointed out that the coldness, hardness, and whiteness of the marble bodies that had been associated with statues from antiquity saved them from any suggestion of the "lower feelings" that could be associated with, for example, John Gibson's (1790–1866) *The Tinted Venus* (*c.* 1851–1856, first shown in 1862). One question that Allen addressed is that if a peacock evolved to attract a peahen, then why should we also find it beautiful? In *The Evolutionist at Large* (1881), Allen explained a series of experiments he performed such as attracting butterflies using colored pieces of paper and he concluded, "when I consider the general continuity of nervous structure throughout the whole animal race, and the exact similarity of the stimulus in each instance, I can hardly doubt that the butterfly really enjoys life somewhat as we enjoy it, though far less vividly."[37] And in "Aesthetic Evolution in Man" (1880) he wrote, "every crow must think its own mate beautiful."[38] Allen believed that beauty was associated with health; he wrote, "The ugly [...] must always be (in the main) the deformed, the aberrant, the weakly, the unnatural, the impotent. The beautiful for every kind must similarly be (in the main) the healthy, the normal, the strong, the perfect, and the parentally sound."[39] This argument, that beauty is a signal of health is still widely discussed today as it provides an explanation for beauty based not on sexual selection but on utility and natural selection. The peacock's tail can then be explained as a massive use of resources designed to demonstrate the owner's superior health and, therefore, fitness.

Peacock and ostrich feathers, sunflowers, and blue china were objects associated with the Aesthetic Movement, and the cult of the Aesthete became a fashionable "pose" associated with refined tastes, but the cult also raised questions of sexual identity, with its unmanly men and manly women, and the symbol of the peacock had also always had a darker side associated with envy and fear.[40] In 1878, Ruskin was suffering bouts of madness and had hallucinations that included the Devil in the form of a peacock,

> I thought I was in a farmyard and that I was impelled by the tyrant Devil to do some fearful wrong, which I strove with all my might and main to resist. But my passionate efforts were to no avail; and every time I did the wrong I heard the voice of the Demon—that is, the peacock—give forth a loud croak of triumph.[41]

In 1882, the year Darwin and Rossetti died, Oscar Wilde engaged in a tour of the US promoting his new aesthetic philosophy. Wilde described the new philosophy as "our English Renaissance" with "its passion for physical beauty, its exclusive attention to form, its seeking for new subjects for poetry, new forms of art, new intellectual and imaginative enjoyments."[42] In Wilde's play *Salome* (1891 French, 1894 English), Herod offers 50 white peacocks to Salome but she refuses and demands the head of Iokanaan (John the Baptist), who describes her as "Daughter of Sodom," alluding to another type of beauty, such as that suggested by Dostoevsky, where procreation is no longer a concern. The play is a profound analysis of the complexities of beauty in the modern world. Although Salome can be seen as a liberated woman who uses the power she possesses through her beauty, she is ultimately frustrated and doomed. Beauty is here seen as sexual desire resulting from the act of looking; however, it results not in procreation, but in frustration and the death of first the Syrian, then the Page, then Iokanaan, and finally Salome herself.

Aubrey Beardsley (1872–1898) often used the peacock and peacock feathers as symbols of decadent beauty. In his final illustration for the published account of *Salome, The Climax,* he reduced the peacock tail to a refined background pattern of overlapping concentric circles (Figure 2.4). In the original sketch that won him the commission, the peacock feathers were boldly stated, but in this final version, they were stylized into a readily recognizable signature motif, set against a large blank area. Beardsley was only 21 years old when he drew these illustrations and their stark message and clear-cut line-block technique increased his fame and notoriety. In *The Climax,* one foreground plant stands upright struggling to survive, but its companion droops flaccidly like a spent force. In the background, a tendril squirms upward, more contorted than a young shoot and more like an animal's tentacle feeling its way towards Salome, who represents the unobtainable object of desire. The picture can be interpreted as the terminal point of a public battle between Ruskin's spiritual beauty, Darwin's scientific beauty, and beauty for its own sake—having now become overwrought, even grotesque, with something of

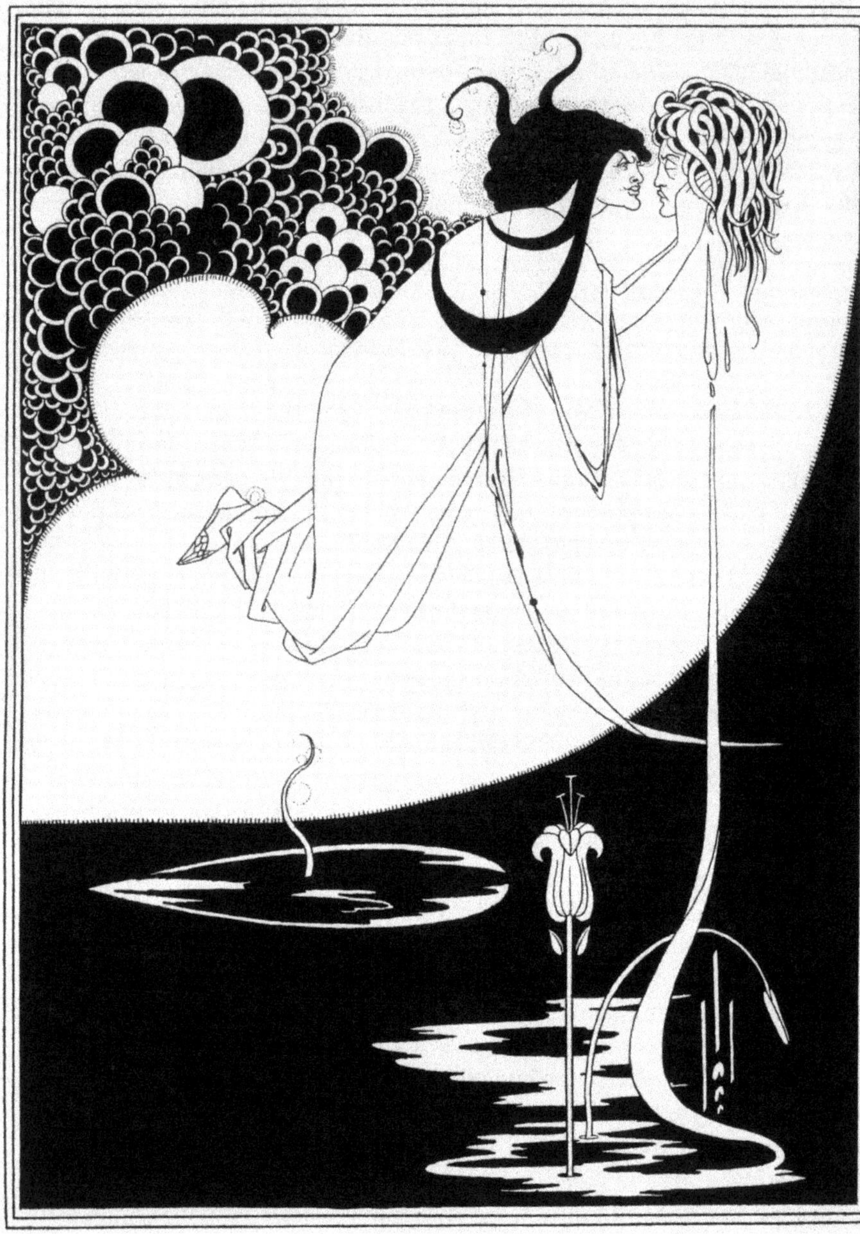

2.4 Aubrey Beardsley, *The Climax*, 1894, line-block and half-tone on Japanese vellum, 34.3 × 27.3 cm, © Victoria and Albert Museum, London the steeliness and danger of present times. Salome, the emancipated woman, looks with unflinching eyes into the dead head of John the Baptist with the cold, calculating look of the modern world.

The conflict between opposing theories is often described using the metaphor of war, for example, as a battleground pitting antagonists against one another—in this case, on one side Ruskin, who continued to stress morality in art, and on the other, the supporters of Aestheticism, where moral subject matter was subordinated to purely aesthetic issues of form and color. But it is clear that such a metaphor is too simplistic, as many factors were involved and Darwinian aesthetics and physiology played a key role in the arguments concerning art and beauty in the 1870s and 1880s.

The beauty of the peacock raises many questions on the journey from Madonna to the daughter of Sodom, including whether beauty is a divine creation or the result of incremental natural laws and whether nature or the artist's individuality is the ultimate source of refined beauty. We have also seen some of the arguments that took place over whether morality must be a fundamental prerequisite of art and whether we possess an inherited recognition of beauty or are born with the mind as a blank slate. The materialistic view of the cause and purpose of beauty brought with it the need to analyze our "lower feelings" and an appreciation that natural beauty often appeared to go hand-in-hand with seduction and deceit.

Notes

1. Kenneth Lantz, *The Dostoevsky Encyclopedia* (Westport, CT and London: Greenwood Press, 2004), 6–7.
2. Fyodor Dostoevsky, *The Brothers Karamazov*, trans. David Magarshack (Baltimore, MD and London: Penguin Books, 1958), 1,123.
3. Duke of Argyll, *The Reign of Law* (London: Alexander Strahan, 1867).
4. Strictly speaking, a peacock's large feathers are part of its train not its tail, but I will continue to use Darwin's term.
5. Charles Darwin to Asa Gray, April 3, 1860, *Darwin Correspondence Project*, http://www.darwinproject.ac.uk/entry-2743.
6. Charles Darwin to Alfred Wallace, June 15, 1864, *Darwin Correspondence Project*. http://www.darwinproject.ac.uk/entry-4535.
7. The term *lek* was first used in 1871 and is used to describe a gathering of males of certain animal species for the purposes of competitive mating display. For a summary of recent ideas concerning sexual selection, see Helena Cronin, *The Ant and the Peacock: Altruism and Sexual Selection from Darwin to Today* (Cambridge: Cambridge University Press, 1991) and Geoffrey Miller, *The Mating Mind: How Sexual Choice Shaped the Evolution of Human Nature* (London: Heinemann, 2000).
8. William Michael Rossetti (1829–1919) quoted by Griselda Pollock, *Vision and Difference: Feminism, Femininity and Histories of Art* (London: Routledge, 2003), 124.
9. Giovanni Boccaccio, *Decameron*, trans. by J.M. Rigg (1903), http://www.brown.edu/Departments/Italian_Studies/dweb/texts/. The final sentence starts, "So she, who had lain with eight men, in all, perhaps, ten thousand times, was bedded with him as a virgin, and made him believe that a virgin she was, and lived long and happily with him as his queen."

10. Stanley Weintraub, *Edward the Caresser: The Man Who Would Be Edward VII*, Part 2 (New York: Free Press, 2001), 39.
11. This would not be the only time Leighton associated beautiful women with peacock feathers. His *Girl Feeding Peacocks* (c. 1863, private collection) was reviewed by W.M. Rossetti as exemplifying "beauty for beauty's sake," thus recognizing it as an early example of the Aesthetic Movement; W.M. Rossetti, "The Royal Academy Exhibition," *Fraser's Magazine for Town and Country* 67, no. 402 (June, 1863), 790.
12. I am indebted to Mrs H.E. Alexander, Director of the Fan Museum, Greenwich, for the information about this nineteenth-century fan. The literary association is to the Elizabethan period, when feather fans were in fashion and the classical association to their use by Roman ladies, as Leighton would have known.
13. Charles Francis Badini created a full set of codes that he explained in *Fanology or the Ladies' Conversation Fan* (London: Robert Clarke, 1797).
14. Charles Darwin, *The Descent of Man, and Selection in Relation to Sex*, 1st edn (London: John Murray, 1871), 1, 273, and John Hunter, *Essays and Observations*, ed. Richard Owen (London: John van Voorst, 1861), 1, 194.
15. Darwin, *Descent*, 1st edn, 2, 122.
16. Darwin, *Descent*, 1st edn, 1, 356, 374.
17. John Ruskin, *The Works of John Ruskin*, ed. E.T. Cook and Alexander Wedderburn (London: George Allen, 1903–1912), 4, 143.
18. Ruskin, *Works of Ruskin*, 22, 246. Briareus was a giant monster in Greek legend with a hundred hands and cephalopoda are molluscs, including the octopus. Ruskin is making fun of what, if Darwinism is true, he suggests is the arbitrary nature of why we have one head and two arms and legs.
19. Ruskin, *Works of Ruskin*, 23, 894.
20. Ruskin, *Works of Ruskin*, 25, 36.
21. Ruskin, *Works of Ruskin*, 20, 372–3, endnotes 7–11.
22. Ruskin, *Works of Ruskin*, 20, 373.
23. Ruskin, *Works of Ruskin*, 22, 194–5.
24. Charles Darwin to William Darwin Fox, July 7, 1837, *Darwin Correspondence Project*, http://www.darwinproject.ac.uk/entry-364. Fox was a close friend and a second cousin and Darwin, in a typically modest comment, added he was being vain and never expected his geology to be worthy of consideration by men such as Charles Lyell.
25. Ruskin, *Works of Ruskin*, 26, xx, and Jonathan Smith, *Charles Darwin and Victorian Visual Culture* (Cambridge: Cambridge University Press, 2009), 277–9.
26. Victor Alexander Ernest Garth Marshall (1841–1928) to Charles Darwin, September 7, 1879, *Darwin Correspondence Project*, http://www.darwinproject.ac.uk/entry-12220.
27. Ruskin, *Works of Ruskin*, 26, 336.
28. Smith, 282.
29. Russell Lande, "Models of Speciation by Sexual Selection on Polygenic Traits," *Proceedings of the National Academy of Sciences of the United States of America* 78, no. 6 (June 1, 1981), 3,721–5.
30. Strictly speaking, variation is undirected rather than random, as crossover and inversion involve reusing existing genetic sequences. Mutations are random but occur rarely and most are neutral or deleterious.
31. St George Mivart, *On the Genesis of Species* (London: Macmillan & Co., 1871), 37.
32. Charles Darwin, *The Variation of Animals and Plants under Domestication* (London: John Murray, 1868), 1, 290–91.
33. "Fashionable Avicide," *Punch, or the London Charivari*, December 14, 1867, 239.
34. Robyn Asleson, *Albert Moore* (London: Phaidon Press, 2000), 125–6.
35. John Ruskin, *Modern Painters* (1906; reprint Whitefish, MT: Kessinger Publishing, 2005), 1,24, and Grant Allen, *Physiological Aesthetics* (1877; reprint n.p: Elibron Classics, 2005).

36. Allen, *Physiological Aesthetics*, 39.

37. Grant Allen, *The Evolutionist at Large* (London, Chatto & Windus, 1881), 159.

38. Grant Allen, "Aesthetic Evolution in Man," *Mind* 5, 20 (October, 1880), 445–64.

39. Allen, "Aesthetic Evolution in Man," 449.

40. An early reference to the peacock is *The Aberdeen Bestiary*, folio 59v, 60r, 60v, 60r c. 1200: "The peacock, as Isidore says, gets its name from the sound of its cry. For when it starts, unexpectedly, to give its cry, it produces sudden fear in its hearers. The peacock is called *pavo*, therefore, from *pavor*, fear, since its cry produces fear in those who hear it." http://www.abdn.ac.uk/bestiary/translat/60v.hti.

41. Ruskin, *Works of Ruskin*, 38, 172.

42. The text of Oscar Wilde's lecture "The English Renaissance of Art" has been recreated and is available at: http://www.wilde-online.info/the-english-renaissance-of-art.html. The quotation is from the second paragraph.

3

Art's "contest with nature": Darwin, Haeckel, and the scientific art history of Alois Riegl

Marsha Morton

The Viennese art historian Alois Riegl adopted the Hegelian concept that every cultural product, whether in the field of art, philosophy, science, law or religion, was a historical document that instantiated prevailing beliefs unique to its era. Riegl was fully aware that his own ideas were equally captive to the reigning *Zeitgeist*, noting that all "scholarship takes its direction…from the contemporary intellectual atmosphere."[1] Accordingly, recent postmodernist art historians, sympathetic to Riegl's message of aesthetic relativism, interdisciplinary contextualization, and reception theory, have documented affinities between his theories and late nineteenth-century directions in philosophy, psychology, and history. Scholars have not, however, attempted to position his writings within his scientific milieu, which was dominated by responses to Darwin. This omission has continued, despite the fact that Riegl defined his age as part of the "natural scientific world view" and his own work as "scientific art history."[2] Indeed, contemporary art criticism by writers such as playwright Hermann Bahr and Franz Wickhoff, Riegl's colleague at the University of Vienna, was permeated with evolutionary references. As an article in the 1899 issue of the *Jugendstil* magazine *Pan* proclaimed: "only to the extent that a modern aesthetics is founded on this theory [Darwinism] will the striving of our age be satisfied."[3]

This chapter will demonstrate that Riegl's work evidences an absorption of evolutionary theory in its methodology, language, content (man's interaction with the environment, forces of competition, free will and determinism), and even ways of conceptualizing problems and processes. Cultural Darwinism in Austria and Germany during the 1890s, however, was no longer associated with strict positivism and materialism, but with modifications that injected spirituality, imagination, and ethics into nature. This shift was exemplified by the prominent University of Jena zoologist Ernst Haeckel, whose earlier materialist orientation had transitioned into monistic pantheism, but with roots in the writings of Alexander von Humboldt that could also be found in Darwin. Like Haeckel, Riegl came to embrace a concept of evolution that was compatible with systems of order, community, and the unity ensured by natural laws. "Today's worldview," Riegl wrote in 1898, conceives of "nature as tangible manifestations of an eternal world spirit."[4]

That "view" was evident in the imagery of the Symbolist Vienna Secession group and the writings of Jung Wien during the 1890s, the first Austrian art to incorporate a Darwinian perspective. Darwin had been embraced by scientists and the educated public in Austria since the 1860s, when the Liberal Party came to power. He was the recipient of numerous honorary awards during the 1860s and 1870s, and the only living scientist whose sculpted bust was included on the façade of the Museum of Natural History, designed by Gottfried Semper in the early 1870s.[5] Haeckel was equally welcomed in Vienna, where he occasionally lectured and was offered a teaching position at the university in 1870. Detailed reviews of publications by Darwin and Haeckel appeared frequently in the Viennese press, especially the liberal *Neue Freie Presse*, which became the central media organ of Darwinism, and whose science editor, Theodor Hertzka, favored a "monist world view" as early as 1875.[6]

Darwinian traces in *Questions of Style*

Riegl's first major book, *Questions of Style* (1893), cited Darwin and Haeckel in its opening pages, though in a manner critical to many Darwin loyalists. The book was written to disprove one application of evolutionary theory to aesthetics (by the Semperians), and replace it with a different, but still Darwinian, theory of artistic origins and development. In so doing, Riegl allied himself with an emerging generation doubtful about an exclusive allegiance to positivism.

Questions of Style was a polemic against the followers of the prominent German and Viennese architect Gottfried Semper, such as the archaeologists Alexander Conze and Reinhard Kekulé, and the conviction that artistic products were determined solely by materials, tools, and functions, with art arising spontaneously and indigenously in imitation of, initially, weaving patterns and, later, forms in nature. Semper, although he had chastised Darwin for limiting the role of free will in creative acts, nonetheless presented an evolutionary argument maintaining that all ornamental motifs and architectural forms could be traced back to a limited number of *Urmotive* (original motifs) related to basic crafts and their materials (weaving, woodwork, stone, ceramics).[7] Riegl's objective in *Questions of Style* was to dethrone this mechanistic–technical narrative (premised on the belief that art represented a developmentally higher intellectual phase of mental activity than utilitarian craft skills and therefore could not have been practiced by early man), and replace it with an account that foregrounded an immanent creative drive present in human nature from the beginning. He did so, however, in ways that were also related to the methods and rhetoric of Darwin and Haeckel. "I aim," he wrote, "to refute the exaggerated claims made for technology in regard to the earliest forms of art. I do not intend to dispute the value and significance of the materialist movement in art of the last 20 years, or even less to criticize the theory of Darwin and his followers."[8]

Riegl inverted the order of the "technical–materialist" theory, asserting that "desire" preceded and determined tools and techniques in a "continuous struggle with material."[9] The need to create art was conceived of as a biological imperative, and described in a scientific and anthropological language. The imagery of the recently discovered Dordogne cave paintings and carvings, for example, was attributed to survival: "The animals depicted…were either a source of food or a source of danger."[10] So too "elemental human drives" accounted for the urge to fill space through drawing and decoration. According to Riegl, the tattoos and jewelry of primitive tribes proved that the desire to adorn the body was greater than the need to cover it with woven garments.[11] In a similar vein, Darwin cited an anonymous "English philosopher" who maintained that "clothes were first made for ornament and not for warmth."[12] Riegl based his "counter argument" (to the overly materialist Semperians) on the assumption that all humans—even in a primordial state—have the capacity "to take pleasure in formal beauty."[13] Darwin also discussed the aesthetic preoccupations of primitive man and quoted Alexander von Humboldt on the "fertile imagination" of "painted nations."[14]

Empirical analysis, morphology, and genealogy

Riegl opposed the idea of art originating in fortuitous, independent, and unrelated acts of imitation. Instead, *Questions of Style* articulated a theory of artistic creation, based on Darwin's "community of descent," as a diachronic "continuous, coherent development" of culturally inherited forms adapted to suit new tastes and to solve new problems of successive generations and societies.[15] His purpose was twofold: to trace the "genetic relationship" governed by the "all-pervasive laws of causality" of a family of antique plant ornaments, whose single progenitor was the Egyptian lotus; and to "examine how the blossoms, buds, and leaves of each style…are connected as they fill up the decorative field."[16] Variations on three different planar projections (frontal, profile, and combined palmette) and their incorporation in the geometric vegetal scroll produced "progeny" in the Mycenaean tendril, Greek acanthus, and Orientalist and Medieval arabesque.[17]

According to Riegl, his approach differed from other art historical research, which established "critical distinctions." "This book," he wrote, "tends decidedly in the opposite direction. Things once considered to have nothing in common will be connected and related from a unified perspective."[18] Riegl's understanding of evolution as an ordered process through kinships of forms and laws—themes he developed in later writings—has a specifically Germanic flavor, where biologists concentrated on the historical dimensions of their subject through genealogy and morphology.[19] Haeckel proclaimed the core tenet of his first book—and magnum opus—*Generelle Morphologie* (1866) to be "the absolute unity of Nature…ruled by the all powerful and unchanging causal laws."[20] This led him to create his famous "family tree" (*Stammbaum*) diagrams, avoided by Darwin, that map patterns of descent.[21]

Riegl's methodological focus on the empirical analysis of objects reflected his training at the Austrian Institute for Historical Research, where scientific procedures had been imported to the fields of history and philosophy. His close scrutiny and description of visual evidence, which at times sounds like the writings of a botanist examining a specimen, are employed to uncover key moments of evolutionary change, such as the birth of the three-dimensional naturalizing acanthus (on capitals) (Figure 3.2) from a two-dimensional corner-filling half-palmette (Figure 3.1), itself the descendent of an Egyptian palmette lotus.[22] The crucial transitional link was discovered to be the hull or pod motif, rotated in space. Riegl's sleuthing for formal clues suggests parallels with the profoundly visual orientation of Haeckel, who was an amateur artist. His work in the Goethean tradition of morphology, which Darwin praised as the "very soul" of natural history, led Haeckel to create the doctrine of promorphology, or the morphology of basic stereometric type-forms which underlay all of nature, both organic and inorganic.[23] These were illustrated in *Generelle Morphologie* (Figure 3.3). His theories had been inspired by the structural similarities between the crystal and the siliceous shell of the radiolaria (Figure 3.4), and enabled him to fill gaps (between organic and inorganic) in the Darwinian system. The increasingly complex symmetrical relations of species allowed Haeckel to literally picture evolutionary development. Both Haeckel and Riegl shared the belief that formal patterns embodied larger truths. As Olaf Breidbach has stated, "For Haeckel, nature reveals itself in its forms and in the order of its forms which is the result of a historical process…What nature is, is visible on its surface."[24]

Riegl's familiarity with these aspects of Haeckel's writings is evident in several of his publications. For Riegl, all art was "inextricably tied to nature," whether it was realistic, when forms are borrowed from nature, or geometric (stylized), when artists create according to natural laws, which he identified as symmetry and rhythm. These, he noted, "are apparent in the natural form of humans, animals, plants, and crystals as well."[25] While principles of crystallization and symmetry are central to nineteenth-century theories of architecture and design, as, for example, in texts by Semper and Burckhardt,

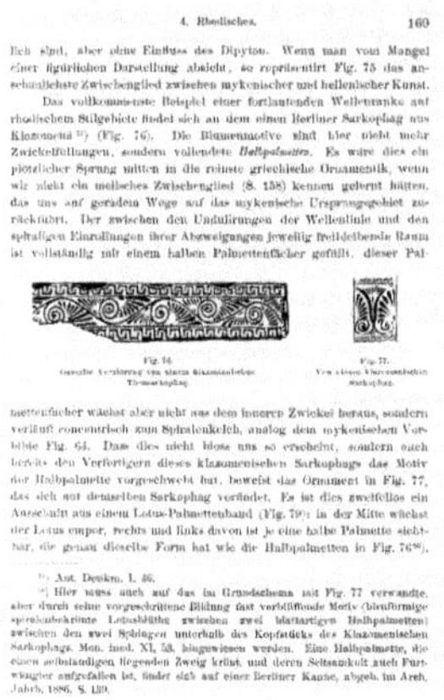

4. Rhodisches. 160

lich sind, aber ohne Einfluss des Dipylon. Wenn man vom Mangel einer figürlichen Darstellung absieht, so repräsentirt *Fig.* 75 das anschaulichste Zwischenglied zwischen mykenischer und hellenischer Kunst.

Das vollkommenste Beispiel einer fortlaufenden Wellenranke auf rhodischem Stilgebiete findet sich an dem einen Berliner Sarkophag aus Klazomenä[77]) (Fig. 76). Die Blumenmotive sind hier nicht mehr Zwickelfüllungen, sondern vollendete *Halbpalmetten*. Es wäre dies ein plötzlicher Sprung mitten in die reinste griechische Ornamentik, wenn wir nicht ein melisches Zwischenglied (S. 158) kennen gelernt hätten, das uns auf geradem Wege auf das mykenische Ursprungsgebiet zurückführt. Der zwischen den Undulirungen der Wellenlinie und den spiraligen Einrollungen ihrer Abzweigungen jeweilig freibleibende Raum ist vollständig mit einem halben Palmettenfächer gefüllt. dieser Palmettenfächer wächst aber nicht aus dem inneren Zwickel heraus, sondern verläuft concentrisch zum Spiralenkelch, analog dem mykenischen Vorbilde Fig. 64. Dass dies nicht bloss uns so erscheint, sondern auch bereits den Verfertigern dieses klazomenischen Sarkophags das Motiv der Halbpalmette vorgeschwebt hat, beweist das Ornament in Fig. 77, das sich auf demselben Sarkophag vorfindet. Es ist dies zweifellos ein Ausschnitt aus einem Lotus-Palmettenband (Fig. 79): in der Mitte wächst der Lotus empor, rechts und links davon ist je eine halbe Palmette sichtbar, die genau dieselbe Form hat wie die Halbpalmetten in Fig. 76[78]).

[77]) Ant. Denkm. I. 46.

[78]) Hier muss auch auf das im Grundschema mit Fig. 77 verwandte, aber durch seine vorgeschrittene Bildung fast verblüffende Motiv (birnförmige spiralenbekrönte Lotusblüthe zwischen zwei blattartigen Halbpalmetten) zwischen den zwei Sphingen unterhalb des Kopfstücks des Klazomenischen Sarkophags, Mon. ined. XI, 53, hingewiesen werden. Eine Halbpalmette, die einen selbständigen liegenden Zweig krönt, und deren Seltsamkeit auch Furtwängler aufgefallen ist, findet sich auf einer Berliner Kanne, abgeb. im Arch. Jahrb. 1886, S. 139.

3.1 Alois Riegl, *Stilfragen* (1893), Ornamentation with corner-filling palmette fans

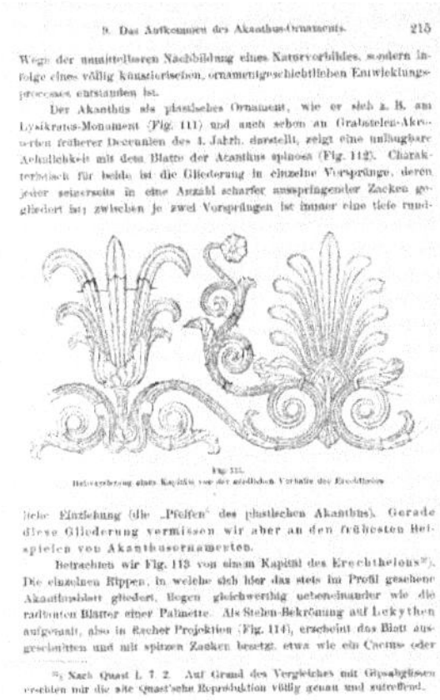

9. Das Aufkommen des Akanthus-Ornaments. 215

Wege der unmittelbaren Nachbildung eines Naturvorbildes, sondern infolge eines völlig künstlerischen, ornamentgeschichtlichen Entwicklungsprocesses entstanden ist.

Der Akanthus als plastisches Ornament, wie er sich z. B. am Lysikrates-Monument (*Fig.* 111) und auch schon an Grabstelen-Akroterien früherer Decennien des 4. Jahrh. darstellt, zeigt eine unläugbare Aehnlichkeit mit dem Blatte der Acanthus spinosa (Fig. 112). Charakteristisch für beide ist die Gliederung in einzelne Vorsprünge, deren jeder seinerseits in eine Anzahl scharfer ausspringender Zacken gegliedert ist; zwischen je zwei Vorsprüngen ist immer eine tiefe rundliche Einziehung (die „Pfeifen" des plastischen Akanthus). Gerade diese Gliederung vermissen wir aber an den frühesten Beispielen von Akanthusornamenten.

Betrachten wir Fig. 113 von einem Kapitäl des Erechtheions[87]). Die einzelnen Rippen, in welche sich hier das stets im Profil gesehene Akanthusblatt gliedert, liegen gleichwerthig nebeneinander wie die radiirten Blätter einer Palmette. Als Stelen-Bekrönung auf Lekythen aufgemalt, also in flacher Projektion (Fig. 114), erscheint das Blatt ausgeschnitten und mit spitzen Zacken besetzt, etwa wie ein Cactus- oder

[87]) Nach Quast I. 7. 2. Auf Grund des Vergleiches mit Gipsabgüssen erschien mir die alte Quast'sche Reproduktion völlig genau und zutreffend.

3.2 Alois Riegl, *Stilfragen* (1893), Ornamentation with full and half acanthus leaves (from the north porch of the Erechtheion)

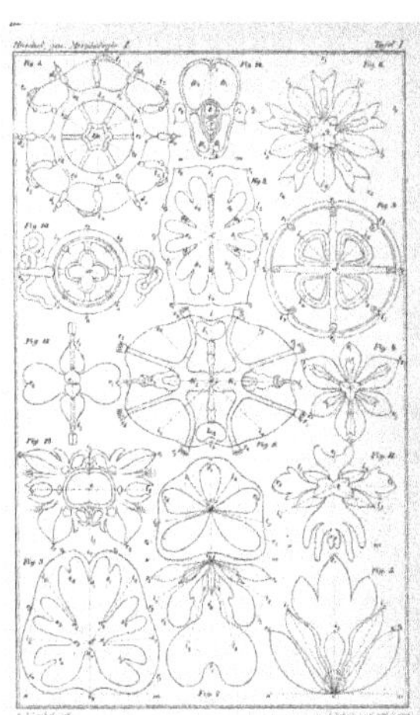

3.3 Ernst Haeckel, "Basic Heteropolar Forms," Plate I, *Generelle Morphologie*, 1866

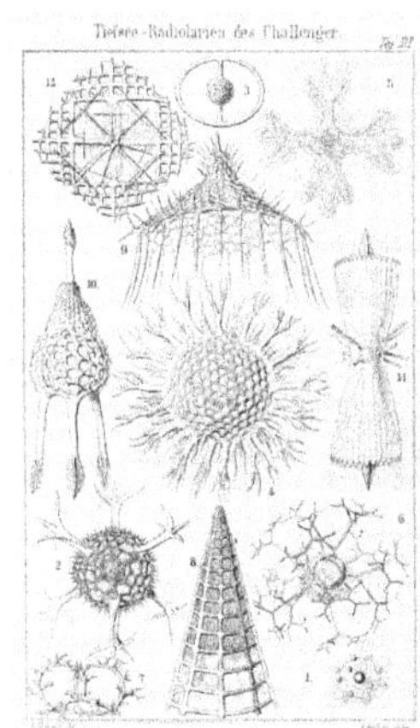

3.4 Ernst Haeckel, "Deep-sea Radiolarians from the *Challenger* Expedition," in *Natürliche Schöfungsgeschichte*, 1889, 8th edition

Riegl's extensive analysis of these concepts in his book *Historical Grammar of the Visual Arts* (1897–1899), and his conjunction of them with organic–inorganic unity implicates Haeckel. Echoes of Haeckel's statements can be found in Riegl's descriptions of the "stereometric and planimetric" symmetry of crystals and Riegl's contention that "nature gives shape to raw matter according to principles of crystallization, which is grounded in the law of symmetry: that is, the equal lateral distribution of even the tiniest molecules along an ideal central axis."[26] Riegl even provides detailed explanations for why symmetry is less evident in organic bodies, and attributes their curved shapes to edges worn away "through the struggle for survival."[27] Ultimately, Riegl categorizes all artistic form as fluctuating between the two poles of crystalline / inorganic and organic forms.[28]

Evolution, variation, natural selection and the *Kunstwollen*

Evolution provided the conceptual structure for nearly all of Riegl's writings. He applied it to ancient art in *Late Roman Art Industry* (1901), to a comprehensive survey of Western art in the *Historical Grammar of the Visual Arts*, to a specific period of Netherlandish Baroque painting in *Dutch Group Portraiture* (1902), and even to the discipline of art history in the essay "Art History and Universal History" (1898). Such developmental frameworks were standard fare in the nineteenth century, however, with Hegel as the most famous exemplar. The difficult question is whether more specific Darwinian parallels can be discovered, particularly regarding the mechanisms of variation and natural selection. I would argue that while no exact equivalencies exist, affinities inevitably emerge, given their—as Riegl would have said—shared "intellectual atmosphere." Riegl's omission of the English word "evolution"—except in his 1902 essay "Late Roman or Oriental?," where he opposed it to "revolution"—should not be taken as a rejection of Darwinian affiliations.[29] As Paul Weindling has noted, most German scientists, including Haeckel, avoided the foreign word in favor of *Entwicklung* or *Entwicklungsgeschichte*, which carried associations of both the new theory of evolution and the old meaning of development.[30] Like Darwin, and even more, Haeckel, Riegl conceived of evolution as "constant progress" with no "unbridgeable gaps" or irreversible "regression."[31] Just as Darwin discussed human evolution through "slow and interrupted steps" over vast periods of time, with occasional reversions to "some of the characters of an early progenitor," so, too, did Riegl define artistic development as unfolding "over thousands of years" and marked by "spurts of progress mixed with moments of regression" and "archaic survivals."[32] As with Darwin, dead ends do occur, though rarely.[33]

The Darwinian evolutionary process can be imagined as a dramatic production, in which new variations of existing species audition on the environmental stage and natural selection evaluates the performances and chooses the winners. As to the causes of those variations, however, Darwin acknowledged that he had as yet incomplete information.[34] By comparison, Riegl's scenario hinges on variations of artistic forms, with the selection made by the *Kunstwollen*, a principle meaning "that which wills art," or, as Joseph Masheck recently translated the phrase, "what art wants to do."[35] Essentially, it represents the aesthetic desires of the collective social body, or artistic *Zeitgeist*. As in biological evolution, variations occur seemingly at random and causes remain undiscovered. While Riegl was convinced that innovative stylistic "seeds" mature slowly over time, the initial change is often "unexpected" and "sudden," new adaptations just "appear," motifs are chosen by "accident," and new directions are introduced sometimes "unintentionally," as with the asymmetrical handle ornamentation of an Attic black figure amphora.[36] In either case—scientific or artistic—variations can be tied to changing conditions of life, both environmental and societal, and are aided by transference and diffusion.

In both stagings of evolution, individuals (whether plants and animals for Darwin or artists for Riegl) are engaged in a dynamic with larger supra-individual controlling forces: natural selection or the *Kunstwollen*, both functioning as the "engine" (Robert Richards) or "motor" (Michael Gubser) of the entire process.[37] In this regard, they are the internal autonomous dynamic of the telic structure within nature or art. Riegl makes it clear that only those artistic variations which conform to the prevailing *Kunstwollen* will be adopted; others remain dormant until a future period when a different *Kunstwollen* will recognize the potential of a variation and allow it to be developed. Darwin and Riegl both resorted to anthropomorphic metaphors to describe these principles, thereby establishing even closer relationships through the implied degree of agency. "Natural selection," Darwin wrote, "is daily and hourly scrutinizing…the slightest variations; rejecting those that are bad, preserving and adding up all that are good; silently and insensibly working, whenever and wherever opportunity offers."[38] Likewise, Riegl defined the *Kunstwollen* as "definite and purposeful," "a determining factor with the phenomenon of evolution," "turning to its own ends every conceivable practical purpose and raw material, and always and of its own accord selecting the most appropriate technique for the intended work of art."[39] In reality, it is only the *Kunstwollen* that is both "controlling" and "generating," as Otto Pächt observed in a lecture at the Courtauld Institute in 1962.[40] The *Kunstwollen* can, and does, exert its will through the artist. While Riegl does allow for some artistic initiative, "it really is art," as Jacqueline Jung has written, and "not (or only rarely) the artist that is in charge."[41]

The *Kunstwollen*, of course, differs completely from natural selection in that it is a cultural imperative and fuels artistic change through its own evolution. For Riegl, this was the basis for his rejection of timeless aesthetic norms in favor of relative standards of beauty. Art was judged according to the period's own values and goals, and according to its contribution to the genealogical tree. He argued, for example, that the stylistic weaknesses of Egyptian art were offset by the fact that they "were the first to achieve" a higher purpose in art: the expression of "perceptions, feelings, and ideas that had nothing in common with pure delight in beauty alone."[42] Most famously, he rehabilitated late Roman art, deemed ugly "from the viewpoint of modern taste," as an essential transitional link in the chain leading to modernism through its liberation of space and introduction of unclassical features.[43] Like Darwin, Riegl was a passionate defender of the underdog, regarding earlier progenitors as "ennobled" by their position within evolutionary kinship.[44] As many have observed, Riegl's championing of aesthetic relativism accorded with the goals of the Vienna Secession, whose motto was "to every age its art."[45] His views were also in the spirit of Darwin, whose world travels had made him cognizant of the subjectivity of beauty: "We must not judge the tastes of distinct species by a uniform standard nor by the standard of man's state." "It is certainly not true," Darwin opined, "that there is in the mind of man any universal standard of beauty with respect to the human body."[46] A related issue inspired the famous defense in 1900 of Klimt's controversial University of Vienna mural *Philosophy* by Riegl's colleague Franz Wickhoff. Incensed by criticisms of the painting as "ugly," Wickhoff mounted an evolutionary justification, insisting that such aesthetic concepts were socially constructed and changed with each generation. As an example, he noted the primitive meaning of beauty, no longer operative, as based on the survival imperative.[47]

Art in competition with nature

Riegl may have jettisoned idealized principles of classical beauty, but he did not abandon the notion of art as an improvement of nature. His definition of art and its goals, like his methodology, are a product of the age of Darwin. Artistic purpose, as theorized by Riegl, was not to imitate but to compete with nature towards the realization of an enhanced image as mankind would like it to be, which would facilitate his happiness (the goal of all human will). The need for such corrections was predicated on Riegl's acceptance of Darwinian realities regarding the brutality and unpredictability of nature, in the face of which art seeks to offer a more comforting vision. Riegl wrote in the 1899 version of *The Historical Grammar of the Visual Arts*: "Man yearns incessantly for harmony. He sees this harmony constantly disrupted and threatened by things and phenomena of nature that exist in a state of perpetual struggle, both with one another and with humanity. Consequently, man creates a vision in his art that frees him from nature's perpetual instability...he seeks to bring order to the apparent chaos, to push aside those raw random occurrences to which he is otherwise subject and vulnerable." "The creation of art," Riegl concluded, "is a contest with nature with the aim of bringing to expression a harmonious worldview."[48] These ideas are consistent with directions in late nineteenth-century German thought which sought to offset the alienating "dark side of nature," characterized by "a relentless war of all against all," as Haeckel phrased it in his 1892 book *Monism as Connecting Religion and Science*.[49] This was accomplished either by promoting a spiritualized and aestheticized image of nature, as Haeckel and Wilhelm Bölsche did, or by offering art as a "metaphysical solace" to Darwin's "true but deadly" doctrine in the manner of Nietzsche and, earlier, Schopenhauer.[50]

The course of art as governed by the *Kunstwollen* thus documents shifting strategies for visualizing harmony, depending on how mankind perceives his relationship with his natural environment at a given historical moment. This development, as constructed by Riegl, itself records a reassuring societal evolution from a primal "survival of the fittest" mentality in antiquity to an enlightened modern world that seeks interrelationships and community. In effect, Riegl mapped out the "struggle against the struggle for existence," advocated by Darwinian sympathizers Friedrich Lange and Ludwig Büchner as the path to universal harmony.[51] Primitive and antique man, according to Riegl, confronted a dangerous "perpetual battle of natural forces," beyond his understanding or control. Art, therefore, responded to an arbitrary and threatening nature and created fixed images, isolated and circumscribed, that established order, regularity, and symmetry—properties inherently associated with harmony. Borrowing from the terminology of the sculptor Adolf Hildebrand, Riegl described this spatial orientation as the haptic or near view, embodied by sculpture, which emphasized individual parts to the exclusion of movement and space. Throughout antiquity, Riegl observed, the "right of the strongest" prevailed as natural law: "Man reigned as the strongest power over all nature...the strong person ruled over the weaker; and the gods...over man and nature." Capitulation to the dominant power ensured peaceful coexistence; in art this translated into a desire for a "physically improved nature" in Greek art that was "perfect [and] enduring."[52] During the Middle Ages, humanity found reassurance through religious beliefs and art accordingly sought to portray a spiritually enhanced vision of nature.

The modern age, which Riegl defined as post-Renaissance, adheres to a "natural scientific worldview" that is nature-centric, in that man ranks "only as much as any other natural thing."[53] Harmony is now derived from the reassuring predictability of "eternally fixed and immutable" causal laws through which all things in nature are united in shared "obedience" and are comprehensible to man.[54] The purpose of modern art, whose essential expression Riegl and Haeckel both located in landscape, is to represent those causal relationships in which the artist "does not isolate objects but shows them within their surroundings."[55] This can be done most effectively through "optical vision," or the panoramic distant view, enabling a comprehensive grasp of objects in a unified spatial and atmospheric setting, observed in adherence to the scientific laws of light, reflection, and color. According to Riegl, "pure optical appearances are the unembellished causal relations of things."[56] Riegl had in mind both Dutch landscape painting and Impressionism. As Nicola Gauld has recently shown, this same trajectory, from isolated individual to interactive group, was followed in illustrations for surveys of natural history, from the Comte de Buffon to Alfred Brehm.[57] Similar beliefs informed the literary criticism of the Jung Wien writer Hermann Bahr, who, as early as 1891, in "Zur Kritik der Kritik," identified the distinguishing feature of modern art as an awareness of Darwinian theory, a cognizance "of all things in a ceaseless flow, and the insight into the connection of all things, in the interdependence of one with the other in an endless chain of being."[58] By the turn of the century he, like Riegl, would emphasize unity and advocate for an artistic integration of *Sinne und Seele*, which he believed had been accomplished by Klimt.[59]

Riegl found the appeal of causal laws to twentieth-century viewers to be also evident in the attractiveness of ruins, whose "age value" evidenced "the purely natural cycle of becoming and passing away."[60] This phrase, *Werden und Vergehen*, had been the title of a popular book on natural history by Darwinist Carus Sterne in 1876, and was in Germany a catch phrase for principles of evolutionary science. As Riegl and Haeckel both observed, it signaled the silver lining of evolution: death countered by "the incessant renewal of life."[61] For Riegl, sensitivity to the rhythms of nature had also produced another related feature of modern culture, the "protection of animals and of the environments," which was advocated by the *Lebensreform* movement.[62]

Quite obviously, Riegl's focus on "the universal connectedness of all natural phenomena," with causality as the guarantor of order, derived from Darwin and Haeckel, just as much as his awareness of nature as the setting for "endless fighting, destruction and discord."[63] Darwin's famous "tangled bank" passage emphasized the mutual interdependency of diverse species "all…produced by laws acting around us."[64] Haeckel went much further, reiterating in each book his allegiance to a unified harmony of nature. Not surprisingly, he established the science of ecology as a sub-discipline of evolution.[65] In *General Morphology*, he urged the study of "life forms in their collective entirety, in their general and continuous mutual interrelationships," while in *The Evolution of Man* he defined his monist philosophy as the assertion that "everywhere the phenomena of human life, as well as those of external nature, are under the control of fixed and unalterable laws, that there is everywhere a necessary causal connection between phenomena, and that, accordingly, the whole knowable universe forms one undivided whole."[66] According to Haeckel, such laws—particularly the "supreme law, the law of substance," which was also cited by Riegl ("the indestructibility of matter and the conservation of energy")—mitigated the chance factor in the evolutionary process.[67]

The significance of the 'big picture' also informed ideas by Haeckel and Riegl about their own disciplines. Riegl advocated for a new third generation of art historians who would move beyond the "cult of individual facts" associated with the positivist-influenced second generation (his teachers) to explore "shared…unifying, generalizable qualities."[68] So, too, Haeckel urged his fellow scientists to blend observation with philosophical interpretation. In *The History of Creation* he criticized "exact empirical naturalists" who "live in the strange delusion that it is possible to erect the edifice of natural science out of mere facts, without their philosophic connection."[69]

Spiritualized nature

The modern *Kunstwollen*, however, in its quest to improve upon nature, sought to reinstate a sacred dimension lost in the era of positivism. To the unifying function of scientific laws and the distant view in art, Riegl added the unifying condition of *Stimmung*, a term suggesting mood or atmosphere, and implying a somewhat mystical state of oneness ("being in tune") with nature. In his 1899 essay, "Die Stimmung als Inhalt der modernen Kunst" ("Die Stimmung as the Content of Modern Art"), Riegl contextualized it within his concept of artistic purpose as "the notion of law and order over chaos, harmony over dissonance, [and] tranquility over movement."[70] Riegl wrote this article in the literary form of the *Plauderie*, adopted frequently by popular science authors such as Bölsche, which employs a first-person narrative and conversationally "paints" a verbal picture. In Riegl's essay, the reader is invited to join the narrator atop a mountain range (literally the distant view) as he gazes over a panoramic space into the "infinite all" and experiences the "spirituality, stillness, and harmony," for which modern man longs.[71] From this vantage point, whose description calls to mind the painted vistas of German Romantic Caspar David Friedrich, as well as native Alpine ranges such as Rudolf von Alt's *The Large Fir in Bad Gastein*, the narrator's reveries are disturbed by a Darwinian tableau of predator and prey: the report of a rifle and fleeing deer. It is these realities of the "near view"—confirmation of the endless "cycle of life and death" and the brutal "struggle for existence"—that must be countered by modern art's metaphorical "view from the mountain top," which satisfies the human desire for "unity" with "an intangible world soul."[72] As instances, Riegl cited, among others, the landscapes of Max Liebermann and Max Klinger (Figure 3.5).

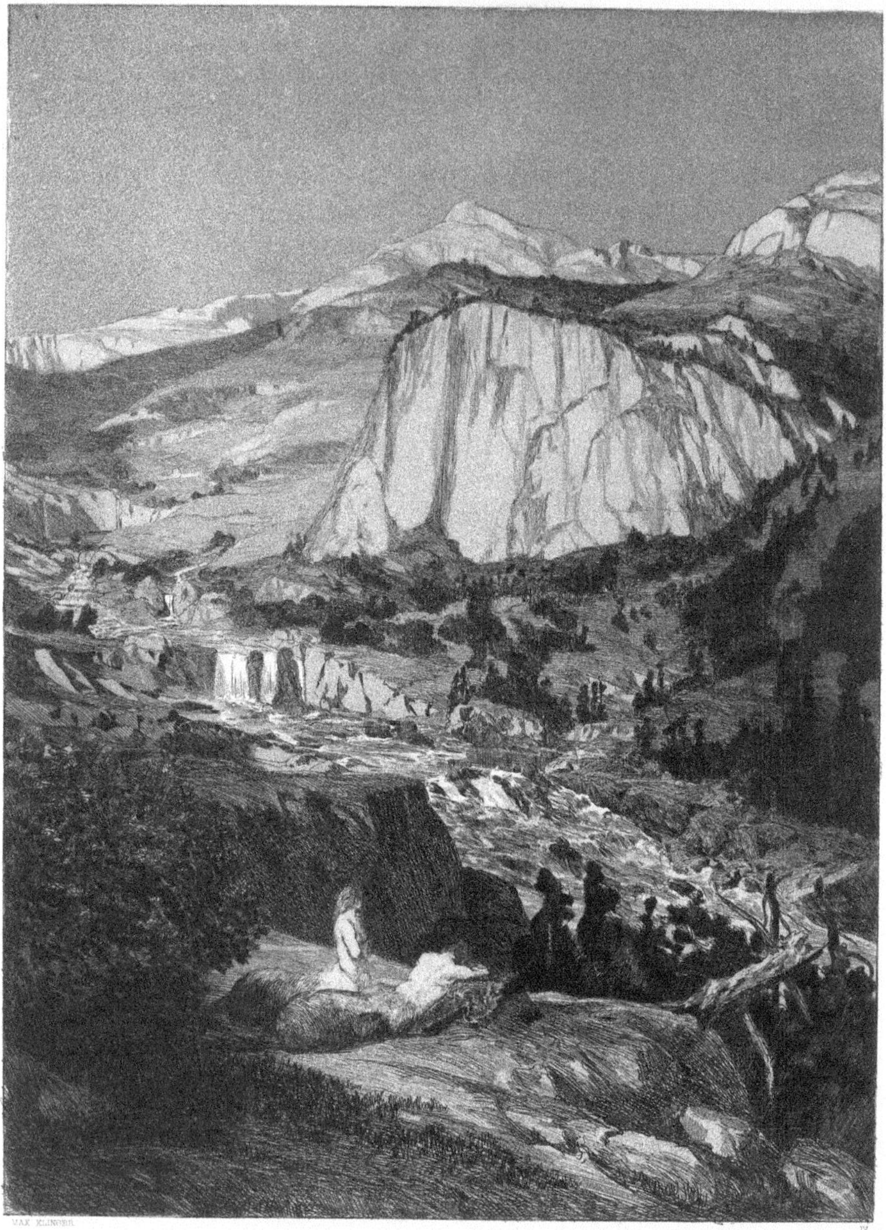

3.5 Max Klinger, *Moonlight,* from the series *Intermezzi,* Opus IV, etching and aquatint, 1881. Gift of the Estate of Mrs Edward Robinson, 1952 (52.594.27). The Metropolitan Museum of Art, New York, NY, US. Image copyright © The Metropolitan Museum of Art / Art Resource, NY

The state of silent contemplation necessary to experience *Stimmung* was merged with the concept of attention in Riegl's later book, *Dutch Group Portraiture*. Netherlandish paintings, according to Riegl, competed with nature by picturing an ideal world in which the drive to dominate yielded to selflessness, facilitating communication and connection with others, and participation in the "all-embracing world soul." "Stimmung," Riegl wrote, "is close to religious devotion (*Andacht*.)"[73]

Riegl's alliance of scientific progress and pantheistic unity was consistent with Haeckel's doctrine of monism and Humboldt's romantic vision of nature as a cosmos, which Darwin so admired. "Everywhere," Humboldt wrote, "the mind is penetrated by the same sense of the grandeur and vast expanse of nature, revealing to the soul, by a mysterious inspiration, the existence of laws that regulate the forces of the universe."[74] Haeckel's monism was predicated on a hylozoic universe in which spirit was immanent in matter, and the "human soul" was subsumed in "the all-embracing world soul."[75] Riegl, in a nod to Haeckel's discoveries, noted in "Die Stimmung" that the latest scientific knowledge provided explanations regarding "the causal relations of all appearances in nature—the physical and recently also the spiritual".[76]

The evolution of Riegl's thought, as has been frequently noticed, adhered to the same trajectory he perceived in art and art history: from the formalist observation of individual facts in *Questions of Style* to an interpretation of interrelationships, whether spiritual or psychological, derived from material evidence. In his journey from being a positivist (albeit opposed to mechanistic arguments) to a positivist with mystical leanings, he also followed the path of many German scientists and artists at the end of the century as they fully digested and responded to the ramifications of Darwinian theory. Art's utopian agenda, as theorized by Riegl, confirms the eternal validity of the sentiments written by the poet Robert Bly a century later: "Only we with our opposable thumbs want Heaven to be and God to come."

Notes

1. Alois Riegl, *Late Roman Art Industry*, trans. Rolf Winkes (Rome: Giorgio Bretschneider, 1985), 6.
2. Riegl, *Historical Grammar of the Visual Arts*, trans. Jacqueline E. Jung (New York: Zone Books, 2004), 337. This reference occurs in the second (1899) version of the manuscript. For the German text see Riegl, *Historische Grammatik der bildenden Künste*, eds Karl M. Swoboda and Otto Pächt (Graz-Köln: Hermann Böhlaus, 1966).
3. E. Freiherr von Bodenhausen, "Entwicklungslehre und Aesthetik," *Pan* 4, (1899): 236, translated in Kurt Bayertz, "Biology and Beauty: Science and Aesthetics in *Fin-de-Siècle* Germany," in *Fin de Siècle and Its Legacy*, eds Mikuláš and Roy Porter (Cambridge: Cambridge University Press, 1990), 279.
4. Riegl, *Historical Grammar*, 186 (*Historische Grammatik*, 128.)
5. Christa Riedl-Dorn, *Das Haus der Wunder: Zur Geschichte des Naturhistorischen Museums in Wien* (Vienna: Verlag Holzhausen, 1998), 187. The most detailed study of Darwinism in Austrian culture is Werner Michler, *Darwinismus und Literatur: naturwissenschaftliche und literarische Intelligenz in Österreich, 1859–1914* (Vienna: Böhlau, 1999).

6. Michler, 75. See also his essay in *Welträtsel und Lebenswunder: Ernst Haeckel—Werk, Wirkung und Folgen*, eds Erna Aescht, Gerhard Aubrecht, Erika Krausse and Franz Speta, exh. cat. (Linz: Oberösterreichisches Landesmuseum, 1998). For information concerning the impact of Haeckel on Gustav Klimt, Alfred Kubin, and Hermann Bahr, see my essay "Nature and Soul: Austrian Responses to Ernst Haeckel's Evolutionary Monism," in *Darwin: Art and the Search for Origins*, ed. Pamela Kort, exh. cat. (Frankfurt: Schirn Kunsthalle, 2009), 126–41.

7. Semper's statements on Darwin are contained in his 1869 lecture in Zurich, "On Architectural Styles." It is published in Gottfried Semper, *The Four Elements of Architecture and Other Writings*, trans. Harry Francis Mallgrave and Wolfgang Herrmann (Cambridge: Cambridge University Press, 1989), 268. For a more detailed discussion of Riegl and Semper, see, among other sources, Mallgrave, *Gottfried Semper: Architect of the Nineteenth Century* (New Haven: Yale University Press, 1996), 374–81, and Margaret Iversen, *Alois Riegl: Art History and Theory* (Cambridge, Mass. and London: The MIT Press, 1993), 22–7.

8. Riegl, *Problems of Style (Stilfragen)*, trans. Evelyn Kain (Princeton, NJ: Princeton University Press, 1992), 23.

9. Ibid., 33.

10. Ibid., 31.

11. Ibid.

12. Charles Darwin, *The Descent of Man, and Selection in Relation to Sex* (New York: Modern Library, n.d.), 881.

13. Riegl, *Problems of Style*, 40.

14. Darwin, 882.

15. Ibid., 411 and Riegl, *Problems of Style*, 9.

16. Riegl, *Problems of Style*, 228. The most comprehensive discussion of this development is in Margaret Olin, *Forms of Representation in Alois Riegl's Theory of Art* (University Park, PA.: Pennsylvania State University Press, 1992), Chapter 4, "Questions about Style," and Olin, "Self-Representation: Resemblance and Convention in Two Nineteenth-century Theories of Architecture and the Decorative Arts," *Zeitschrift für Kunstgeschichte* 49, 3 (1986), 376–97.

17. Ibid., 184.

18. Ibid., 12.

19. Paul Weindling, *Darwinism and Social Darwinism in Germany: The Contribution of the Cell Biologist Oscar Hertwig (1849–1922)* (Stuttgart: Gustav Fischer, 1991), 28. As Paul Weindling has observed, "Haeckel set greater store on causality than Darwin did; he believed that the discovery of the correct historical succession of organisms constituted a mechanical explanation of morphology." Weindling, "Ernst Haeckel, Darwinismus, and the Secularization of Nature," in *History, Humanity and Evolution: Essays for John C. Greene*, ed. James R. Moore (Cambridge: Cambridge University Press, 1989), 318.

20. Ernst Haeckel, *Generelle Morphologie der Organismen*, vol. 1 (Berlin: Georg Reimer, 1866), 447. This book has never been translated into English.

21. For these differences between Darwin and Haeckel, see Julia Voss, *Darwins Bilder: Ansichten der Evolutionstheorie 1837–1874* (Frankfurt: Fischer Taschenbuch Verlag, 2007), 147–63.

22. Olin, *Forms of Representation*, 77–9.

23. Darwin, *On the Origin of Species* (New York: Modern Library, n.d.), 333.

24. Olaf Breidbach, *Art Forms from the Ocean*, trans. Stephen Telfer (Munich: Prestel, 2005), 18.

25. Riegl, *Problems of Style*, 15.

26. Riegl, *Historical Grammar*, 51, 123; see also 343–7.

27. Ibid., 127–8 and 133.

28. Ibid., 133.

29. Riegl, "Late Roman or Oriental?," trans. by Peter Wortsman in *German Essays on Art History*, ed. Gert Schiff, The German Library, vol. 79 (New York: Continuum Press, 1988), 177.

30. Weindling, *Darwinism and Social Darwinism*, 30.

31. Riegl, "Late Roman or Oriental?," 15, and *Problems of Style*, 81.

32. Darwin, *Descent of Man*, 511, *Origin of Species* 117, and Riegl, *Historical Grammar*, 251, and "Late Roman of Oriental?," 17.

33. Riegl, *Problems of Style*, 70.

34. Darwin, *Origin of Species*, 101. In *Origin* he retracted his earlier use of the word "chance," dismissing it as "a wholly incorrect expression" which nonetheless "serves to acknowledge plainly our ignorance of the cause of each particular variation."

35. Joseph Masheck, "The Vital Skin: Riegl, the Maori, and Loos," in *Framing Formalism: Riegl's Work*, in the series *Critical Voices in Art, Theory and Culture*, ed. Saul Ostrow (Amsterdam: G+B Arts International, 2001), 164.

36. Citations are listed in the order of the quoted words: Riegl, *Historical Grammar*, 60, *Problems of Style*, 181, *Historical Grammar*, 95, 137, and 133. Movement in Egyptian art is presented as an example of unintended change.

37. Robert Richards, *The Tragic Sense of Life: Ernst Haeckel and the Struggle over Evolutionary Thought* (Chicago: University of Chicago Press, 2008), 147, and Michael Gubser, *Time's Visible Surface: Alois Riegl and the Discourse on History and Temporality in Fin-de-Siècle Vienna* (Detroit: Wayne State University Press, 2006), 151.

38. Darwin, *Origin of Species*, 66.

39. Riegl, *Late Roman Art Industry* 9, 94–5.

40. Otto Pächt, "Art Historians and Art Critics—vi: Alois Riegl," *Burlington Magazine* 105 (May 1963), 190.

41. Jacqueline Jung, "Preface," in Riegl, *Historical Grammar* 42. A similar observation was made earlier by Michael Ann Holly when she wrote that, "the artist initiates change but is also initiated." *Panofsky and the Foundations of Art History* (Ithaca and London: Cornell University Press, 1984), 76.

42. Riegl, *Problems of Style*, 81.

43. Riegl, *Late Roman Art Industry*, 11.

44. Darwin, *Origin of Species*, 41.

45. Riegl cited this in *The Modern Cult of Monuments: Its Character and Its Origin*, trans. Kurt W. Forster and Diane Ghirardo in *Oppositions* 25 (Fall, 1982), 48.

46. Darwin, *Descent of Man*, quoted by Diana Donald, "Introduction," in *Endless Forms: Charles Darwin, Natural Science and the Visual Arts*, eds Diana Donald and Jane Munro, exh. cat. (Cambridge: Fitzwilliam Museum and New Haven, CT: Yale Center for British Art, 2009), 20; and Darwin, *Descent of Man*, 890.

47. Franz Wickoff, "Was ist Hässlich?," May 15, 1900 lecture, excerpted in *Gegen Klimt*, ed. Hermann Bahr (Vienna: J. Eisenstein, 1903), 31–2. For a discussion of this text in relation to Riegl, see Gubser, 121, and Iversen, Chapter 3, "The Aesthetics of Disintegration."

48. Riegl, *Historical Grammar*, 299–300. The German word Riegl used for contest is "Wettschaffen." He writes of nature's "beständigem Kampf" and its "rohen Zufall." Riegl, *Historische Grammatik*, 216–17 (see note 2).

49. Haeckel, *Monism as Connecting Religion and Science*, trans. J. Gilchrist (Glos., UK: Dodo Press, n.d.), 73. The book was reviewed by Haeckel-enthusiast Benjamin Carneri, "Glaubensbekenntnis eines Naturforschers," in the *Neue Freie Presse*, 31 March 1893, 4.

50. Bayertz, 283. Friedrich Nietzsche, "On the Uses and Disadvantages of History for Life," in *Untimely Meditations*, trans. R.J. Hollingdale (Cambridge: Cambridge University Press, 1990), 112. For a more detailed discussion of this direction, see Andreas W. Daum, "Das versöhnende Element in der neuen Weltanschauung: Entwicklungsoptimismus, Naturästhetik und Harmoniedenken im populärwissenschaftlichen diskurs der Naturkunde um 1900" in *Vom Weltbildwandle zur Weltanschauungsanalyse: Krisenwahrnehmung und Krisenbewältigung um 1900*, eds V. Drehsen and W. Sparn (Berlin: Akademie Verlag, 1996), 203–15.

51. Friedrich Lange, *Die Arbeiterfrage in Ihre Bedeutung für Gegenwart und Zukunft* (Duisberg: W. Falk & Volmer, 1865; 2nd edn 1870), 368, quoted in Weikart, *Social Darwinism*, 89, and Ludwig Büchner, *Man in the Past, Present, and Future*, trans. W.S. Dallas (London: Asher & Co., 1872), 175.

52 Riegl, *Historical Grammar*, 307.

53. Ibid., 339–40.

54. Ibid., 337.

55. Ibid., 339. In *The Modern Cult of Monuments*, Riegl noted that, "The development of the *Kunstwollen* is tending toward an integration of the individual art object with its surroundings," 32.

56. Ibid., 387.

57. Nicola Gauld, "'What is Meant by this System?' Charles Darwin and the Visual Reordering of Nature," in *Endless Forms*, 119–40.

58. Hermann Bahr, "Zur Kritik der Kritik," in *Zur Kritik der Modern* (Zurich: Verlags-Magazin, 1890–1891), 250.

59. The phrase is from Bahr, "Rothe Bäume," in *Renaissance* (Berlin: S. Fischer, 1897), 201. A devoted follower of Haeckel after the turn of the century, Bahr wrote essays during the late 1890s, when he was an editor for the Secession journal, *Ver Sacrum*, that demonstrate a new belief in the importance of a unified matter and spirit. Discussing Leopold Andrian's novel *The Garden of Knowledge*, he stated that, "the sense of life is to be sought not in details but in the whole, the unity of all things." (*Renaissance*, 44.) For further information, see Donald Daviau, *Understanding Herman Bahr* (St Ingbert: Röhrig Universitätsverlag, 2002), 113.

60. Riegl, *The Modern Cult of Monuments*, 32.

61. Ibid.

62. Ibid.

63. Ibid., 361. Riegl, "Die Stimmung als Inhalt der modernen Kunst," in Riegl, *Gesammelte Aufsätze* (Vienna: WUV Universitätsverlag, 1996), 27. The essay was originally published in *Graphische Künste*, XXII, 1899, 47–54.

64. Darwin, *Origin of Species*, 373.

65. For more on Haeckel and ecology see Richards, 144.

66. Haeckel, *Generelle Morphologie*, vol. 2, 240, translated in Richards, 143. Haeckel, *The Evolution of Man* (London: C. Kegan Paul & Co., 1879), 455.

67. Haeckel, *The Riddle of the Universe*, trans. Joseph McCabe (1899; Amherst, NY: Prometheus Books, 1992) 276. Riegl, *Historical Grammar*, 337.

68. Riegl, "Naturwerk und Kunstwerk I," 1901, in *Gesammelte Aufsätze*, 1996, 51, and "Kunstgeschichte und Universalgeschichte," 1898, in *Gesammelte Aufsätze* (Berlin: Gebr. Mann Verlag, 1995), 7, trans.. and discussed in Gubser 129.

69. Haeckel, *The History of Creation*, vol. 1, trans. E. Ray Lancaster (New York: D. Appleton, 1876), 349.

70. Riegl, "Die Stimmung," 28.

71. Ibid., 27.

72. Ibid., 28. For a discussion of this essay in relation to contemporary art see Olin, 122–7.

73. Riegl, "Die Stimmung," 36.

74. Alexander von Humboldt, *Kosmos*, 1 (Stuttgart: Cotta'scher Verlag, 1845–1858), 25, quoted in Robert Richards, *The Romantic Conception of Life: Science and Philosophy in the Age of Goethe* (Chicago: University of Chicago Press, 2002), 520.

75. Haeckel, *Monism*, 16.

76. Riegl, "Die Stimmung," 31.

4

Cultural selection and the shape of time

Larry Silver

Genealogy is not an historical narrative, but has the essential function of renewing our perception of the…system as in an X-ray…serving to make perceptible the articulation of functional elements of a given system in the present.

(Frederic Jameson, *The Political Unconscious*)

Each species is a biological experiment, and there is no way to predict, as far as an incipient species is concerned, whether the new niche it enters is a dead end or the entrance into a large new adaptive zone.

(Ernst Mayr, *This is Biology*)

Modern viewers of art take for granted conventional pictorial categories such as landscape or genre painting (scenes of daily life and ordinary people). Yet these categories have their own origins and early histories, connected to the media upon which such scenes commonly appear (easel paintings and prints). In fact, the rise of now traditional pictorial genres and their media coincided, and both flourished for the first time in sixteenth-century Antwerp, the most volatile commercial and financial center in Europe and the home of the first permanent open art market, located at the City Bourse.[1] This chapter will examine this important moment in the "evolution" of Dutch art and will suggest that financial and social pressures gave rise to the lineages that seem so normative today, which thus can be understood better from the perspective of selection. This kind of analysis, a form of "cultural Darwinism," renders emergent pictorial categories not as inevitable, but as the result of quite specific circumstances that proved to be historically viable.

In late medieval Flemish painting saints—including those found in landscapes—were common. Landscape emerges as a category of its own during the early sixteenth century within the depiction of saints in the wilderness. But origins are notoriously difficult to pinpoint. At what moment does the subject of a saint with landscape background (traditional) become a painting of a landscape with a saint as a detail (emergent)? That is precisely the transition that occurred during the first quarter of the sixteenth century in Antwerp, particularly in the influential oeuvre of Joachim Patinir, designated already in 1521 by the visiting Albrecht Dürer in his diary as "the good landscape painter."[2] Modern scholars have agreed in assigning Patinir (d. 1524) an innovator's role in the formulation of landscape as a genre, despite the presence of saints in his settings.

The term "landscape," extending from the time of Dürer's designation to the present day, can seem to imply a coherent, continuous category that begs a history with a moment of origin.[3] While such a trajectory may be cogent, historically and conceptually, we must beware of drawing the often-voiced, anachronistic conclusion that in reaching back to Patinir and crediting him as the first landscape painter, we do not overstep our observations by implying he has inserted a token saint into a scene to bow to pious viewers, when in fact he really wanted to be a landscape painter. This landscape-based teleology underlies Max Friedländer's twentieth-century assertion that Patinir provided the origin for later instances of landscape painting all the way to Impressionism and beyond.[4] In his chapter significantly entitled "The Emancipation of Landscape in the Sixteenth Century," Friedländer's classic text claims that, "landscape painting extricated itself from the altarpiece and stood on its own feet...."

This kind of art history posits historical change as inevitable development, with an implied progress narrative and teleology. It locates a prototype as an anticipation or seed of the mature phenotype, or essential body type. This same kind of thinking informs Erwin Panofsky's discussion of Gothic architecture as the fulfillment of an "ideal type" (in the sense used by Max Weber), towards what he calls a "final solution" of a pictorial or structural problem.[5] In similar fashion, earlier scholars have also construed other periods, such as fifth-century BCE Athens or the—*nomen est omen*—High Renaissance in Rome, as classic or climactic historical moments. When it comes to style, built into such discussions can also be implications that types, like organisms, have "lives" of their own, with youth, maturity, and eventual decline, where the high point of maturity constitutes their "classic" phase or peak accomplishment.[6] Yet "organic" progress narratives or birth–life–death models fail to consider the contingency at any moment of alternate solutions and violate the letter as well as the spirit of Darwin's great insights, which can be useful in considering histories of pictorial categories as well as style.

Artworks of similar kinds can be thought of within a framework of linked solutions to common problems, something recognized by art theorists for some time.[7] Antwerp landscapes of the early sixteenth century, with Patinir as an important source or participant, offer a case study for examining a related emerging series, based on what George Kubler—a specialist in Mesoamerican art—calls "prime objects," whose connectedness can be followed despite shifts of form and meaning through the series over time.

As Friedländer noted correctly, Patinir made hybrid works—works that served simultaneously as religious pictures with saints in them as well as "proto-landscapes." His paintings resulted from both inherited pictorial formulas and a specialization in a recognizable kind of painting, even if executed in collaboration with other artists. Friedländer also noted correctly that Patinir's specialization was tied to a new condition of art-making: instead of individual commissions (as had been the case in the past), within the new open art market in Antwerp, art production resulted from a replication of successful pictorial formulas along with distinctive, personalized imagery for marketing purposes. This phenomenon of recognizable artistic identity led to favorite forms and themes, which in turn stimulated viewer (i.e. consumer) interest. Patinir's signature on his paintings, unusual for the period, suggests awareness of the importance of "brand" identification in successful marketing of his familiar formulas.

Just prior to Patinir, one artist, Hieronymus Bosch (d. 1516), began to sign his pictures, and his distinctive, inventive demonic imagery of Hell scenes or tormented saints in vast landscapes turned out to be vastly popular, thereafter widely imitated in sixteenth-century Antwerp.[8] One prominent Bosch imitator during the artist's early period, especially when it came to his prints, was Pieter Bruegel the Elder (active 1551–1569).[9] Later, Bruegel developed his own distinctive inventions, particularly peasants at labor or leisure in panoramic landscapes.[10] These in turn would be imitated by, among others, his own pair of painter sons, Pieter Bruegel the Younger and Jan Bruegel.[11] Thus was born an artistic family dynasty of painters over multiple generations, each practicing the family business with what can only be termed "trademark" consistency.[12] Indeed, such brand recognition of celebrated leading masters generated its own resulting demand, even for their "knock-offs," pictorial imitations and derivations in an era before copyright. Today we take such typical artistic formulas as commonplace, along with the frequent use of identifying signatures (a "Rothko" is instantly familiar, whether signed or not). Yet both circumstances, pictorial and denominated, along with the increasing use of easel paintings as well as prints, resulted directly from early modern art market conditions.

The importance of an artist's specific identity is evident in the case of Albrecht Dürer (1471–1528), whose name became famous throughout Europe, in part because of the wide distribution of his monogrammed prints, previously the prerogative mainly of engravers, such as the anonymous Master ES (d. 1467), beginning in the mid-fifteenth century.[13] Dürer even instituted a lawsuit in Venice against copies of his work made by Marcantonio Raimondi, though the net result protected his author's mark, but not his compositional designs.[14] In similar fashion, the Antwerp print publisher Hieronymus Cock (who represented Bruegel among others), issued a number of prints, including an early one by Bruegel, which bore (falsely) the name of Bosch for marketing purposes.[15]

Bruegel's own career was stimulated not only by Bosch but also by models from an earlier generation of sixteenth-century German printmakers, who had collectively begun to specialize in peasant subjects.[16] Bruegel's innovation consisted in large part of recasting such peasant subjects from prints to a larger scale and full color presentation in paintings. Other contemporary Antwerp artists, such as Pieter Aertsen (active 1535–1575) even preceded him in this, using peasant imagery in their paintings.[17]

Moreover, Bruegel's successes were not achieved without vigorous contemporary protest and polemic. In particular, poet and painter Lucas de Heere, an ardent follower of an alternate, idealizing and Italy-oriented camp of painters in Antwerp, attacked Bruegel in everything but name in a poem, published in 1565, "Invective against a certain painter who scoffed at the painters of Antwerp."[18] Yet Bruegel's formulas enjoyed a long afterlife, not only through the imitations and variations by his own sons but also in the North Netherlands, especially Haarlem and Amsterdam, well into the early seventeenth century.[19]

Meanwhile, the genre of landscape was emerging through successive steps of formal experimentation—from saints in landscapes to Patinir and Bruegel's own landscapes with peasants to an eventual Dutch idiom of local country landscape scenes. Innovations were followed by replications in both prints and paintings, which linked standard subjects with the emerging pictorial formulas of the landscape genre. At the same time, the names of innovative artists achieved brand-name familiarity and stimulated ongoing demand, even for lesser imitations in the competitive art market.

For a much more fully documented instance of the relation between market pressures and selection and the emergent combination of art forms and subjects, we can consider the modern art form of the Hollywood genre film, which emerged from acknowledged industrial and collaborative artistic production.[20] Film historians tend to seek out the studio promotions of films, as defined in their original marketing, to designate their intended genres, such as Westerns or horror. Yet categories used today also influence the construction of the past. Thus, a film's genre emerges out of an ongoing dynamic, through comparison between present genre constructs and particular past instances. Inevitably, therefore, a tension emerges between retrospective distillation of essential components of a genre and its individual instances in an originating work, which might well lack some basic forms that appear later or else have additional complicating elements or variations within the accepted range of basic conventions.

In the case of movies, this means sometimes that certain groups of films might be retrospectively defined or revalued as a new genre; for instance, "women's films" were anachronistically reclaimed in the 1980s, an era of feminism.[21] Also pertinent is the possible revival of certain genres after a long period of quiescence, as was the case with the Western during the early 1990s.[22]

Thus, the dynamics of production and consumption in Hollywood film remind us that movies offer complex interactions around the artwork—between industry production, audience response(s), and critical or scholarly communities, including historical reconstructions. To define a film genre at any moment results from this complex negotiation, comparing any specific work with its previous models and other contemporary examples, reinforced by the tastes of purchasers and *aficianados*, including critics and theorists.[23]

Current genre theory of Hollywood films suggests that recognition and definition of genres occurs retrospectively and often cannot be noted as a self-aware novelty in the early stages (as in "I am now making the first feminist film"). Film historian Rick Altman notes that later critics seek the origins of what they see as a pattern of works in later films, finding them all to have roots in what was initially mixed imagery (like Patinir's saints in landscapes), to serve a variety of needs for a diverse audience, especially in the open, competitive marketplace of film production and distribution.[24] Thus, genres can encompass historical variants, still firmly located within a changing sequence and within shifting boundaries that in retrospect become increasingly more varied and refined. Part of the process of genre formation, as outlined by Altman, involves taking an existing form and modifying it through the emphasis or addition of some new property, which potentially can become a defining feature of a new, generative subgenre. He uses the case study of the "musical comedy," but the same point is valid for Westerns or any other genre.[25]

This process of "genrification" offers conceptual advantages, inasmuch as it can include retrospective audience communities, including critics, and can remain dynamic as well as historically extended over time. It destabilizes the fixity usually accorded to genres as predictable, prior, and prescriptive rosters of traits and patterns, thus permitting a broader, more inclusive, and potentially shifting, notion for any genre. Additionally, it places the process of genre definition within both particular cultural groups and historical moments. This combination of variability and historicity bears an analogy with forms changing over time—evolution—that also underlies natural history. Yet the retrospective potential for regrouping cultural objects into new narratives as in film (e.g. "women's films") or even in pictorial art is a different matter. And because genre-identity is constructed culturally and refined through human choices and readings, film genre history, as well as art genre history, can also comprise sudden or voluntary change—sometimes even by willed adherence to models or traditions under new conditions, such as revivals of dormant forms and themes in revival movements, known as "renascences" or "neo-" movements.[26] Thus, in comparing cultural objects with natural history there is no necessary extinction; much later revivals are possible in cultural reality.

The very phrase "art history" concerns the study of forms over time. Current thinking about the evolution of art itself offers two main approaches.

The first, more traditional, approach, best theorized in George Kubler's *The Shape of Time* (1962), sees art as a sequence of original creations, Kubler's "prime objects."[27] These objects—often called "key monuments"—are assigned critical weight by art historians, to be inscribed in what has come to be called "the canon." These works give rise to the imitations and knock-offs that are broadly characterized as "influence." The greater the influence, the greater the major monument, or so the argument goes.

The other, more currently fashionable, post-structural approach situates art works as collaborative creations rather than individual initiatives that arise out of a life-and-works biography of a genius who "develops." Such works are conditioned and constrained by cultural and social conventions, not to mention a constellation of materials, markets, and other elements of production that precondition manufacture and distribution. In this view, by contrast, the artwork more closely resembles a published book, whose author and publisher, not to mention distributor, all take part in the process. Perhaps an even more collaborative modern art form would be that very studio motion picture, with participants too numerous to mention. Such a film gives "credits" to a whole host of co-producers and individual technicians as well as to actors and a guiding director, who, in some film theories, is given special credit as an *auteur*, despite the constraints and collaborations inherent in studio production.[28] By this reckoning, the replicable print produced in Antwerp by Hieronymus Cock, designed by Pieter Bruegel or some other master, and then engraved by specialist craftsmen, should be seen as the proper equivalent of a modern book or movie.

Both the Kubler formulation and a contemporary theoretical approach yield particular heuristic benefits, classifying works either by innovative artists or else by success in the competition of the marketplace. The change of forms over time can be compared to theories about evolution in the natural world, keeping in mind Rick Altman's justified warning about how much more interaction (movie) genres can elicit than living organisms: "In the genre world, however, every day is *Jurassic Park* day. Not only are all genres interfertile, but they may at any time be crossed with any genre that ever existed."[29]

Clearly some (visual) forms are going to flourish and find replication and imitation—the traditional measure of what art historians call "influence." Others of these works will fall flat, either from being too costly or too radically inventive to find an audience or else just unattractive to potential buyers for any reason when compared with the competition. What these pictures have in common with the Darwinian world of "nature red in tooth and claw" is precisely that—competition, albeit situated within the marketplace.[30]

Between cultural selection and natural selection, however, other differences remain. Constraints in the biological world often are considerable, if mutable, limited by such facts as heredity and body makeup, or phenotype. A bird's body, for instance, cannot be larger than a certain size in proportion to its wingspan in order to permit efficient flight. In the art world certain pictures, such as altarpieces in churches, cannot be smaller than a certain size in order to be seen across vast spaces (though smaller altarpiece forms were created for personal use or different purposes). As Altman reminds us, however, film genres can indeed intermix, both creatively and arbitrarily, in ways that mutually incompatible species cannot (this very infertility has been identified as the essential definition of the boundaries between species).[31]

Constraints of the open marketplace might be considerable, but they are not absolute. Instead of the long-term conservatism and constancy of genetic material and the physical constraints of living species, pictorial images, by contrast, can be adapted rapidly and at will. Of course, there can often be another kind of pictorial conservatism in the form of audience expectations as well as size, cost, and other market variables.

But when there is a breakthrough, such as the hellish inventions of Bosch or the landscape formulas of Patinir, we can clearly and instantly see the success of the inventor as well as the potential flourishing of a larger circle of followers who imitate those novelties. In other traditions, such as the icon-painting tradition of religious art in Orthodox countries, where replication with relatively little alteration is a long-term trend, innovations are marked only by small, usually individual, but often temporary differences. Indeed, such deliberately conservative situations are often imposed by public authority, such as religion or political power. In such situations, public need, filled by icons or equivalent images of authority, such as pharaoh portraits in ancient Egypt, would clearly be compromised in their purpose by too much formal innovation.

One idea within evolutionary biology is that natural selection can either extend over eons or else be very rapid.[32] For example, bacteria replicate so rapidly and in such quantities that many of them have become penicillin- and drug-resistant against modern medicine in a matter of decades. Sometimes, after an initial period of radical innovation, organisms too can maintain long periods of stasis, an oscillating process that Gould termed "punctuated equilibrium," though this theory has its critics. Artworks, too, often closely connected to politics or religion, settle into long periods of conservative traditionalism after initial periods of experimentation and variety.

Moreover, just as boom and bust cycles affect art sales within the general economic marketplace, so, too, can there be major population expansions and accommodation for greater varieties of species in periods of natural prosperity, such as abundant rainfall, in contrast to droughts and other catastrophic climatic or ecological devastation. The geological history of the Earth is marked by several periods of disastrous extinctions, of which that of the dinosaurs at the end of the Mesozoic Age is the most familiar. Experiments and observations on the controlled environment of the Galapagos by Rosemary and Peter Grant have documented natural selection's effects on morphological traits of the local finches as well as the hybridization within and among species.[33] They discovered that minute differences in beak size and shape could exert enormous differences in survival during lean times, when the population thinned down to the most versatile competitors, yet during periods of ecological accommodation, their overall numbers swelled and attained their greatest variety. Another boom and bust cycle of cross-breeding and intense demand for rarities climaxed in early seventeenth-century Holland during the tulip craze and crash, which even involved artists as speculators, customers, and species portraitists of the coveted flowers.[34]

Eric Jan Sluijter cites documents around 1610, when Dutch painters in Leiden and Amsterdam complained about a flood of inexpensive paintings, especially landscapes from Antwerp, that drove down prices in the northern art market.[35] The Dutch successfully adapted, however, and found their own cheap means to compete in the marketplace, in part through the development of a new process—monochrome painting—as well as etching in preference to engraving. In fact, these means helped produce the next great wave of image innovation, the Haarlem school of landscape in the teens of the seventeenth century, led by Esias and Jan van de Velde, later by Jan van Goyen and Salomon van Ruisdael. Yet even here, those who could not produce rapidly and in quantity had perilous careers; the cautionary tale of Hercules Segers, who died a pauper, despite innovation as a painter and radical experimentation as an etcher shows that not all the major innovators of this generation either thrived or found imitators.[36]

Though his economic explanation has been criticized for being overly mono-causal, Jonathan Israel has suggested that monochrome painting techniques of seventeenth-century Dutch easel paintings arose precisely out of the conditions of an economic downturn during the 1620s as a cost-cutting efficiency by struggling artists.[37] Certainly, the shift to simplified graphic and painting techniques—to etching from engraving, to monochrome tonal washes from slow-drying, multi-layered and detailed Flemish techniques—marked a new kind of simplicity and an adaptation that suggested "directness" as a hallmark of Dutch naturalistic picture-making.[38] While efficiency of production might have been a consideration in the adoption of new painting techniques, particularly for an artist like Jan van Goyen (1596–1656), whose output of low-priced landscapes was prolific, this tonal, basically monochrome imagery was also closely associated with the grey skies and muted colors of the Dutch landscape, so it provided a means to display virtuoso manipulation of limited means for illusionistic effects. Furthermore, this technique provided a striking, local Dutch contrast to the inherited, meticulous Flemish techniques, still practiced by many of the followers of Pieter Bruegel.

While research in this area is still relatively new, it is already clear that such a technical history of workshop procedure and the handling of medium—whether painting or graphics—has its own trajectory. Moreover, abrupt change of technique also forms a competitive strategy, which can draw a vivid, visual distinction between "progressive" Dutch pictures against the "mannered" or "traditional" legacy of earlier (and ongoing) Flemish traditions—in effect fostering a new pictorial taste for a consumer public and a market niche or sales demand among collectors.[39] This change is called process innovation, one major kind of novelty that can make a picture distinctive—the other novelty is product innovation. Both alterations affect art value, according to the late John Michael Montias.[40]

Biological evolution can suggest a number of hypotheses for art historians. For example, one of the main arguments by biologist Ernst Mayr about speciation—the multiplication and eventual diversification of life forms in a setting—is that the natural expansion of types and numbers occurs most readily in a favorable but peripheral environment, such as a tropical island like the Galapagos or even parts of Australia, where biota have been isolated from predators and competitors and can then expand to fill all available niches in the ecosystem.[41] I suggest a parallel cultural situation regarding local marketplace competition in the Low Countries in the sixteenth century. Geographically isolated in northwest Europe and featuring a newly dominant Protestant attitude towards art imagery (which diverged from traditional needs to supply visual works for monarchs and churches), Dutch artists chiefly responded to the successes and failures of their artwork in the marketplace.[42] Those images that were either innovative to a buying audience in acceptable ways or else conventionally competent and readily affordable were able to flourish and find imitations, in which case they became somewhat diverse and varied within the same basic, emerging genre types.

Mutatis mutandis, the early Hollywood film studios with their new cinema technology, their control of movie houses where their films were shown to the public, and their relative geographical isolation in southern California, nevertheless engaged in an intense competition for entertainment time and dollars. At the same time, in this closed cultural environment, individual film studios began to specialize and to acquire their own characteristic "looks" and favorite genres.

What remains significant about the overall analogy to biological evolution by art in a historically situated marketplace is that both models stress variation and selection, with success seen in terms of diversification, longevity, and quantity. Certain art historians, especially Henri Focillon, posited a period style as a kind of organism, with its own "life of forms" and its own growth and change. But we no longer have to use that deterministic biological metaphor for visual form. Instead, we can see changes in artworks and relationships within genres over time—a process that is contingent rather than inevitable or teleological. And if we can abandon the notion of progress for either visual art or life forms (after all, simple bacteria are still a dominant life form on earth), then the historical model of Darwinian evolution has much to teach the historical model of art.

Yet we also need to remember visual objects are cultural artifacts, made by conscious human choices in relation to their models and precedents, as discussed above. Innovation is one option, but so is the desire to return to earlier, previously discarded models (even at a broad cultural level, as the Renaissance and Neoclassicism demonstrate). Defining change as inevitable or charting unfolding, genre characteristics is part of selective hindsight, such as the retrospective critical unification of imagery of markets, kitchens, and flower pieces, originally all discreet form classes in inventories, into a single genre, known afterwards (but not originally) as "still life" (*nature morte* in the French Royal Academy after the later seventeenth century).[43]

Kubler's "shape of time" does still celebrate originality and creativity through "prime objects" exerting an influence over time, but Kubler at least breaks free of (his teacher) Focillon's naïve assumption, harshly critiqued, especially by Gould, that life evolves in progress from simple to complex on an ascending chain of being.[44] Instead, the working model proposed here is that Kubler's ideas of culture history (new genre, subjects, or object types) are not entirely dissimilar to emergent, successfully adapted species that proliferate. And under the conditions of the art marketplace (obviously patronage situations are more constrained), comparisons between competition of open art sales to conditions in natural settings can be made. There, cultural competition and cultural selection determine success; there, numerous chances arise for both innovation and variation, as well as continuity or extinction of any form. This model focuses on forms that hold lasting influence and enables us to construct our own historical chains of succession, as paleontologists trace lineages of particular animal types, such as the modern horse from eohippus in the Pleistocene.

It is also easy to think of artworks that are still considered major, even "great," but that come precisely at the end of a tradition, such as the Isenheim Altarpiece (1515) by Matthias Gotthard Neithard, whose very name seemed lost to history except under the misnomer "Grünewald."[45] Other works often failed to participate in the usual historical art period sequences, only to be revived in a later period; examples include paintings, often seen as eccentric or unique, by such "rediscovered" artists as El Greco and Vermeer.[46]

Biologists have revealed that evolution involves selection between and among whole populations and that changes are neither entirely accidental nor predetermined, but rather the result of survival from wide variation within a conditioning environment. Even unhelpful mutations might still survive alongside advantageous traits. Likewise, no essential form or ideal type prevails across the history of art, nor does change occur predictably or inevitably in a given sequence. In fact, this point is stressed emphatically by another form of historical biology, called cladistics.[47] This particular analysis of forms over time does not attempt to determine how changes occurred, partly because of the "deep time" assumed for most evolutionary changes; as a result it is not much employed by most evolutionary biologists, since it requires a common ancestor and makes crucial the difficult distinction between homology and analogy. Cladistics focuses solely on the points where the branches of the tree of life (*clades*) diverge. It notes that degree of difference does not depend upon how recently a shift occurred. For example, the well-acknowledged kinship between modern birds and a major class of dinosaurs, normally separated in evolutionary history, might form the basis for a cladistic grouping, based on common ancestors. This approach also offers a ready conceptual framework for dealing with the willed, cultural phenomena of either revival or resemblance to establish visual continuities even after long gaps of time, e.g. Renaissance or Neoclassicism in relation to a past ancient world.

Having this flexible model enables us to have our cake of innovation and to eat it too, within the complex interaction of artwork, producer and associates, audience, and market. If we remember that individual artists can produce their own variations and elect to experiment, whether to achieve success or failure in sales and survival, then we can also view their outcomes in the competitive and collaborative world of early modern open art markets for visual art. Thus, we can potentially resolve the tension, even the conflict, between a modernist celebration of individual originality (relevant to a Kubler-based model) and a postmodernist construction of social constraints, collaborations, and contingencies.

What this chapter proposes, then, is a reconciliation between alternate models of analysis by considering works and genres by innovative artists, as well as by the sequences of successful formulas over time in the marketplace. Its working hypothesis is that the market competition and pictorial genres over time is analogous to the situation, well studied since Darwin, of biological evolution by selection. At the same time, as noted earlier, cultural products like paintings and prints are *not* subject to the hereditary constraints or built-in species limits (such as genetics and body size) in the natural world. Thus, we need to take account, not just of pictorial sequences, but also of cultural choices and historically conditioned meanings in the forms and genres that flourished in Antwerp (or Holland or Hollywood). As proposed, the constraints of the marketplace in terms of replicating successful works are considerable. Friedländer already noted the pictorial conservatism of a broad viewing public, and market constraints of size or cost also affect visual objects, especially easel paintings and prints. However, pictures can also be adapted rapidly and at will by innovative artists. When a breakthrough image type is forged by an individual, such as Bosch or Patinir or Bruegel, then the ensuing success of their novelty quickly spills over to a larger circle of imitative followers. Change can be either very rapid or quite gradual, though Antwerp's art market provided a far more volatile, more competitive setting than traditional, centralized societies, whether pharaonic Egypt or the orthodox Byzantine empire.

Notes

1. For this historical moment see Filip Vermeylen, *Painting for the Market: Commercialization of Art in Antwerp's Golden Age* (Turnhout, Belgium: Brepols, 2003), and Dan Ewing, "Marketing Art in Antwerp, 1460–1560: Our Lady's *Pand*," *Art Bulletin* 72 (1990), 558–84.

2. Alejandro Vergara, ed., *Patinir*, exh. cat. (Madrid: Museo del Prado, 2007), esp. 53–5.

3. A recent catalogue exploring these origins is Hans Devisscher, ed., *De uitvinding van het landschap. Von Patinir tot Rubens 1520–1650*, exh. cat. (Antwerpen, Belgium: Koninklijk Museum voor Schone Kunsten, 2004). Broader recent surveys include Boudewijn Bakker, *Landschap en wereldbeeld: van Van Eyck tot Rembrandt* (Bussum: Thoth, 2004), and Nils Büttner, *Landscape Painting: A History* (New York: Abbeville Press Publishers, 2006).

4. Max Friedländer, *Landscape, Portrait, Still-Life* (New York: Schocken Books, 1963), 47.

5. Erwin Panofsky, *Gothic Architecture and Scholasticism* (Cleveland, OH: Meridian Books, 1957), esp. 60–70.

6. Key for this view is Henri Focillon's *The Life of Forms in Art*, 2nd edn (New York: Wittenborn, Schultz, Inc., 1948); also see Paul Frankl, *Das System der Kunstwissenschaft* (Brno / Leipzig: R.M. Rohrer, 1938). For criticism of this outlook, Meyer Schapiro, "Style," *Theory and Philosophy of Art: Style, Artist, and Society* (New York: G. Braziller, Inc., 1994; orig. publ. 1953), 69–81; James Ackerman, "Style," in J.S. Ackerman and Rhys Carpenter, *Art and Archaeology* (Englewood Cliffs, NJ: Prentice-Hall, 1962), 164–86, esp. 170–73.

7. Erwin Panofsky, "The History of Art as a Humanistic Discipline," *Meaning in the Visual Arts* (Garden City, NJ: Doubleday and McClure Company, 1955), 20–21; George Kubler, *The Shape of Time* (New Haven, CT: Yale University Press, 1962), 31–9.

8. Recently, Larry Silver, *Hieronymus Bosch* (New York: Abbeville Press Publishers, 2006), including the later Bosch imitators; also Silver, *Peasant Scenes and Landscapes* (Philadelphia, PA: University of Pennsylvania Press, 2006), 133–60. The rarity of signatures by artists in the fifteenth century is attested by Jacqueline Folie, "Les oeuvres aunthentifiées des primitifs flamands," *Bulletin de l'Institut Royal du Patrimoine artistique* 6 (1963), 183–256.

9. Larry Silver, "Bruegel Translates Bosch," in *The Humor and Wit of Pieter Bruegel the Elder*, ed. Henry Luttikhuizen, exh. cat. (Grand Rapids, MI: Calvin College Art Gallery, 2010), 29–44; Silver, *Pieter Bruegel* (New York: Abbeville Press Publishers, 2011), 137–78.

10. Silver, *Pieter Bruegel*, 93–135, 307–59.

11. Silver, *Pieter Bruegel*, 399–431; *Bruegel—Une dynastie de peintres*, exh. cat. (Brussels: Foundation Europalia International, 1980).

12. Silver, *Peasant Scenes and Landscapes*, 186–207.

13. Janez Höfler, *Der Meister E.S.* (Regensburg: Schnell and Steiner, 2007); on Dürer and the maker's mark, Joseph Koerner, *The Moment of Self-Portraiture in German Renaissance Art* (Chicago: University of Chicago Press, 1993), esp. 203–23; also Koerner, "Albrecht Dürer: A Sixteenth-Century *Influenza*," in *Albrecht Dürer and His Legacy*, ed. Giulia Bartrum, exh. cat. (London: British Museum Press, 2002), 18–38.

14. Lisa Pon, *Raphael, Dürer, and Marcantonio Raimondi* (New Haven, CT: Yale University Press, 2004), 39–43.

15. Henry Luttikhuizen, ed., *The Humor and the Wit of Pieter Bruegel the Elder*, exh. cat. (Grand Rapids, MI: Calvin College Art Gallery, 2010); also Paul Lafond, *The Prints of Hieronymus Bosch*, trans. and ed. Susan Fargo Gilchrist (San Francisco: Alan Wofsy Fine Arts Gallery, 2002), esp. 107–15, nos. 30–38.

16. Alison Stewart, *Before Bruegel. Sebald Beham and the Origins of Peasant Festival Imagery* (Aldershot: Ashgate Publishing, 2008); Walter Gibson, *Pieter Bruegel the Elder: Two Studies* (Lawrence, KS: Spencer Museum of Art, University of Kansas, 2002), 11–52; Hans-Joachim Raupp, *Bauernsatiren* (Niederzier: Lukassen, 1986).

17. Pieter Aertsen, *Nederlands Kunsthistorisch Jaarboek* 40 (1989).

18. Mark Meadow, "Bruegel's Procession to Calvary, Aemulatio and the Space of Vernacular Style," *Nederlands Kunsthistorisch Jaarboek* 47 (1996), 181–82; David Freedberg, "Allusion and Topicality in the Works of Pieter Bruegel," *The Prints of Pieter Bruegel the Elder*, exh. cat. (Tokyo: Curatorial Department of the Bridgestone Museum of Art, 1989), 53–65.

19. Silver, *Peasant Scenes and Landscapes*, 161–85. Bruegel was even erroneously credited during a reprint in Amsterdam in 1612 with the authorship of a print series, now known as the "Small Landscapes," by a still anonymous master designer, who featured 44 more intimate rural scenes of villages and estates. Walter Gibson, *Pleasant Places: The Rustic Landscape from Bruegel to Ruisdael* (Berkeley, CA: University of California Press, 2000), 1–26, 32–6.

20. Out of a large bibliography, see Thomas Schatz, *Hollywood Genres: Formulas, Film-making, and the Studio System* (New York: McGraw-Hill, 1981); Tino Balio, *Grand Design: Hollywood as a Modern Business* (Berkeley, CA: University of California Press, 1995); and especially Rick Altman, *Film / Genre* (London: British Film Institute Publishing, 1999); Steve Neale, *Genre and Hollywood* (London: Routledge, 2000); Christine Gledhill, "Rethinking Genre," in *Reinventing Film Studies*, eds Christine Gledhill and Linda Williams (London: Arnold; New York: Oxford University Press, 2000), 221–43.

21. Annette Kuhn, *Women's Pictures: Feminism and Cinema* (London: Routledge & Kegan Paul, 1982); Mary Ann Doane, *The Desire to Desire: The Woman's Film of the 1940s* (Bloomington, IN: Indiana University Press, 1987); Christine Gledhill, "Mapping the Field," in *Home is Where the Heart Is: Studies in Melodrama and the Woman's Film*, ed. Gledhill (London: British Film Institute, 1987), 5–39. For a revision of the revision, Altman, *Film / Genre*, 72–7.

22. Particularly outstanding in quality if not quantity, with celebrated laureate films by actor–directors Kevin Costner, *Dances with Wolves* (1990) and Clint Eastwood, *Unforgiven* (1992). On the Western as film genre, another long bibliography, see Edward Buscombe, ed., *The BFI Companion to the Western* (New York: Antheneum, 1988); Jim Kitses and Gregg Rickman, eds, *The Western Reader* (New York: Limelight Editions, 1998); compare Kitses's earlier auteurist approach, *Horizons West* (Bloomington, IN: Indiana University Press, 1969). Tag Gallagher, "Shoot-Out at the Genre Corral: Problems in the 'Evolution' of the Western," Barry Grant, ed., *Film Genre Reader* (Austin: University of Texas Press, 1986), 202–16, arguing for a history of the genre in cyclical rather than evolutionary fashion over the course of the century.

23. Critical discourse on artworks begins in Renaissance Italy, but in seventeenth-century Flanders, connoisseurs, known as *liefhebbers*, literally amateurs, or art-lovers, actually enrolled in the Antwerp painters' guild. See Zirka Filipczak, *Picturing Art in Antwerp 1550–1700* (Princeton, NJ: Princeton University Press, 1987), 51–2; Elizabeth Honig, *Painting and the Market in Early Modern Antwerp* (New Haven, CT: Yale University Press, 1998), 202–5. More generally, see Joseph Alsop, *The Rare Art Traditions* (Princeton, NJ: Princeton University Press, 1982).

24. Altman, *Film / Genre*, esp. 30–82, discussing where film genres come from, whether such genres are stable, and whether they are subject to redefinition over time. I wish to acknowledge the foundational influence of this important book, as well as personal dialogue with Professor Altman.

25. Altman, *Film / Genre*, 31–4, 66–8.

26. See the classic study by Erwin Panofsky, *Renaissance and Renascences in Western Art* (Stockholm: Almqvist and Wiksell, 1960).

27. George Kubler, *The Shape of Time* (New Haven, CT: Yale University Press, 1962), 31–82, esp. 39–53.

28. Robert Stam, *Film Theory. An Introduction* (Oxford: Blackwell Publishing, 2000), 83–92; on the structuralist reconsideration of both authorship and genre, 123–30.

29. Altman, *Film / Genre*, 70; he continues, "The 'evolution' of genres is thus far broader in scope than the evolution of species…a record not of the past but of a living geography, of an on-going process."

30. On the Tennyson phrase, Stephen J. Gould, "Red in Tooth and Claw," *Natural History* 101 (November ,1992), 14–23.

31. Ernst Mayr, *This Is Biology* (Cambridge, MA: Belknap Press of Harvard University Press, 1997), 127–34, 311: "A reproductively isolated aggregate of populations which can interbreed with one another because they share the same isolating mechanism."

32. Major sources for this chapter include the works of Ernst Mayr, *Evolution and the Diversity of Life. Selected Essays* (Cambridge, MA: Belknap Press of Harvard University Press, 1976); *This Is Biology*; *What Evolution Is* (New York: Basic Books, 2001). Also Stephen J. Gould, *Wonderful Life* (New York: W.W. Norton & Company, Inc., 1989); *The Structure of Evolutionary Theory* (Cambridge, MA: Belknap Press of Harvard University Press, 2002); Edward O. Wilson, *The Diversity of Life* (New York: W.W. Norton & Company, Inc., 1992); Peter Bowler, *Evolution*, 3rd edn (Berkeley, CA: University of California Press, 2003); Michael Ruse and Joseph Travis, eds, *Evolution. The First Four Billion Years* (Cambridge, MA: Belknap Press of Harvard University Press, 2009). More argumentative—and critical of Gould's concept of "punctuated equilibrium"—Richard Dawkins, *The Blind Watchmaker* (New York: W.W Norton & Company, Inc., 1987).

33. Jonathan Weiner, *The Beak of the Finch. A Story of Evolution in Our Time* (New York: Alfred A. Knopf, Inc., 1994); also David Quammen, *The Song of the Dodo* (New York: Scribner Publishing, 1996), 216–17, 222–30.

34. Ann Goldgar, *Tulipmania: Money, Honor, and Knowledge in the Dutch Golden Age* (Chicago: University of Chicago Press, 2007).

35. Eric Jan Sluijter, "Over Brabantse vodden, economische concurrentie, artistieke wedijver en de groei van de markt voor schilderijen in de eerste decennia van de zeventiende eeuw," *Nederlands Kunsthistorisch Jaarboek* 50 (1999), 113–43.

36. For the period, *Dawn of the Golden Age. Northern Netherlandish Art, 1580–1620*, exh. cat. (Amsterdam: Rijksmuseum, 1993); for Segers, ibid., 316, 672–76, nos. 343–47.

37. Jonathan Israel, "Adjusting to Hard Times: Dutch Art during its Period of Crisis and Restructuring (c. 1621–c. 1645)," *Art History* 20 (1997), 449–76.

38. Lawrence Goedde, "Naturalism as Convention: Subject, Style and Artistic Self-Consciousness in Dutch Landscape," in *Reconsidering Realism: Looking at Seventeenth-Century Dutch Art*, ed. Wayne Franits (Cambridge, MA: Cambridge University Press, 1997), 129–43. I have also benefited from the work of Melanie Gifford, particularly "Jan van Goyen en de techniek van het naturalistische landschap," Christiaan Vogelaar, *Jan van Goyen*, exh. cat. (Leiden: Stedelijk Museum De Lakenhal, 1996), 70–79. Thanks, too, to Dr Gifford for helpful discussions on technique. She also notes that even traditional Flemish painting manner could be used efficiently with less blending of local colors.

39. On appreciation of Jan van Goyen by his contemporaries, such as critics Christian Huygens (1629) and Jan Orlers (1641), as well as his presence in artistic collections, Eric Jan Sluijter, "Jan van Goyen als marktleider, virtuoos en vernieuwer," *Jan van Goyen*, 38–59, whose title characterizes the artist as "market leader, virtuoso, and innovator." See also Mariet Westermann, "*Fray en Leelijck.* Adriaen van de Venne's Invention of Ironic Grisaille," *Nederlands Kunsthistorische Jaarboek* 50 (1999), 221–57, for the case of an artist who "remade himself as a painter, relinquishing his speciality in polychrome painting of aristocratic outings for a novel specialization in abject, even grotesque images of the socially marginal painted in monochrome...and shaped market interest in imagery of lowlife, verbal and pictorial wit, and the grisaille medium," 221.

40. His classic study: J. Michael Montias, "Cost and Value in Seventeenth-Century Dutch Art," *Art History* 10 (1987), 455–66.

41. Most recently, Ruse and Travis, *Evolution*, 93–8, 177–207.

42. Josua Bruyn, "A Turning Point in the History of Dutch Art," in *Dawn of the Golden Age: Northern Netherlandish Art, 1580–1620*, ed. Ger Luijten (Amsterdam: Rijksmuseum, 1994) 112–21.

43. Titia van Leeuwen, "Still-Life Painting in the Netherlands: Historical Facts and Facets," in *Still-Life in the Age of Rembrandt*, ed. Eddy de Jongh, exh. cat. (Auckland: Auckland City Art Gallery, 1982), 39–52; more generally, Alan Chong and Wouter Kloek, *Still-Life Paintings from the Netherlands, 1550–1700*, exh. cat. (Amsterdam: Rijksmuseum, 1999).

44. Ruse and Travis, *Evolution*, 550–57, 744–5; for the history of the idea of a tree of life metaphor, Arthur Lovejoy, *The Great Chain of Being: A Study of the History of an Idea* (Cambridge, MA: Harvard University Press, 1964).

45. Though corrected by twentieth-century research, the misnomer still prevails in art historical literature, e.g. Horst Ziermann, *Matthias Grünewald* (Munich: Prestel Publishing, 2001), or even the most recent exhibition of original works, *Grünewald und seine Zeit*, exh. cat. (Karlsruhe: Staatliche Kunsthalle, 2007).

46. Francis Haskell, *Rediscoveries in Art: Some Aspects of Taste, Fashion, and Collecting in England and France*, 2nd edn (Ithaca, NY: Cornell University Press, 1980).

47. Ruse and Travis, *Evolution*, 141–2, 260–61; Bowler, *Evolution*, 371–74; Henry Gee, *In Search of Deep Time* (New York: Free Press, 1999). I am grateful to Dr Harry Rand for introducing me to this form of phylogenetic analysis.

5

The evolution of culture or the cultural history of the evolutionary concept: epistemological problems at the interface between the two cultures

Sigrid Weigel

Evolution defined as a natural genealogy—the classification of "species in a state of nature" (Dawkins, Darwin)

One of the concepts of cultural evolution that enjoys the greatest popularity today goes back to Richard Dawkins' introduction of the "meme" in analogy to the molecular biological concept of the gene, understood as a unit of genetic information. The purpose of this neologism was to enable scholars to form units that would allow cultural development to be described using the paradigm of evolutionary theories:

> We need a name for the new *replicator*, a noun that conveys the idea of a *unit of cultural transmission*, or a *unit of imitation*...Examples of memes are tunes, ideas, catch-phrases, clothes fashions, ways of making pots or building arches. Just as genes propagate themselves in the gene pool by leaping from body to body via sperms or eggs, so memes *propagate themselves* in the meme pool by leaping *from brain to brain* via a process which, in the broad sense, can be called imitation (emphasis mine).[1]

Problems related to the formation of theories and concepts can best be studied in the arenas in which such concepts are introduced. It is no surprise, then, that this passage from Dawkins' book *The Selfish Gene* (1976) reveals the methodological conditions and limits of his concept. The examples he provides—tunes, ideas, catch-phrases, fashions and production methods in the crafts and architecture—demonstrate that by looking at *units* of cultural evolution, he only includes those cultural skills that are highly conventionalized. From the perspective of evolutionary theory, these are situated along the axis of stability, invariance, and continuity. And even if this axis is extended to include the dimension of change—whether it is change by selection or variation—the use of units to describe cultural evolution necessarily restricts "culture" to its most conventionalized register: rituals, symbolic formula, and institutionalized behavioral patterns—or, as it were, the *canon* of cultural phenomena. This restriction also explains why, in humanities research, the paradigms of evolutionary theory are applied primarily to the analysis of social systems, the development of linguistic signs or tools, and to legal history, ethnology, and cultural anthropology.

An additional problem illustrated by the quoted passage is that the meme, in analogy to the genetic function of self-reproduction and replication, is a medium for propagating culture by means of which cultural evolution gets described metaphorically as a process of transmission and reproduction within the paradigm of *procreation*. However, if cultural processes are reduced to the operations identified by Dawkins—replication, transmission, and imitation—only a quite limited perspective is opened on the broad field of inquiry into the history of culture. This field includes an enormous variety of methodologies and concepts in order to study the actual production and tradition of cultural practices and meanings. After all, replication, transmission, and imitation belong to the cultural register of *techné*, *poiesis*, *mimesis*, and *aisthesis*. As regards their history, we can make out a near endless repertoire of figures and concepts that interact with the history of writing systems, techniques, knowledge cultures, institutions, and their extremely different historical, national, and aesthetic practices. Most importantly, the intergenerational transmission of these processes does not proceed "from brain to brain," as Dawkins writes, but through media, language and symbolism. It would go beyond the scope of this chapter to examine the broad research field of cultural history—to discuss, for example, the diverse reproduction, duplication, citation and simulation processes that developed in the twentieth century alone on the basis of photography, film, new printing technologies, and new digital imaging technologies. These introductory remarks are merely meant to illustrate that the complex field of research that confronts us here cannot be grasped by a concept from evolutionary theory. In this respect Dawkins' meme takes us in, for his concept of cultural evolution requires an analysis of the material and symbolic systems through which memes are transmitted along the path between two brains. Without studying the media through which fashion or architectural production methods, for example, are communicated, Dawkins' paradigm of cultural evolution remains mired in a metaphorical attempt to describe the processes of cultural transmission using the image of biological reproduction.

One would, perhaps, not need to discuss Dawkins' idea in greater depth, nor Edward O. Wilson's efforts to develop this idea into a "complete theory of gene–culture co-evolution," if not for the fact that the appearance of the meme in Dawkins' book reveals a structural epistemological problem associated with the model of cultural evolution.[2] This problem remains in effect even if evolution is not defined in the classic (Darwinian) sense of a theory of both descent and inheritance based on derivation, modification, selection, and higher development. It remains in effect even if evolution is liberated from its ties to the illusion of progress or is understood as the differentiation of structures and the specialization of functions—a view that can be found in the more recent system-theoretical versions of evolutionary theory (see below). This problem arises from one of the necessary conditions for evolutionary theories—that they must establish units, "species" or even "systems" as their subjects and arrange these units in a continuum along the temporal axis in order to describe changes of these units and to explain the laws governing these changes and sometimes also their directedness. In the definitional language of *Geschichtliche Grundbegriffe*, this idea is expressed as follows: "An identical, persistent subject forms the basis of change; this subject can be a supra-individual formation, a figure of the 'objective spirit.'"[3]

Evolutionary theories are always inscribed with ideas about classification—that is, with decisions about distinct entities. In the field of biology, these units appear as a natural given in the form of organisms. The fascination that is apparently exerted by Darwin's model of evolution and that has inspired others to imitate, develop, and transfer it as the history of species to other fields can be assumed to originate in the combination of *genealogical* and *taxonomic* operations that render the order of things capable of representation as a family tree or a natural system. The establishment of a *natural genealogy* can be seen as one of the most important elements of evolutionary theory. For example, in Chapter 13 of the *Origin of Species* (1859), which touches upon the problems of classification, Darwin repeatedly emphasizes the connection between the genealogical and the natural: "I believe that the *arrangement* of the groups within each class, in due subordination and relation to the other groups, must be strictly genealogical in order to be natural."[4] Whereas this passage makes the validity of a "natural" taxonomy dependent on the condition of a genealogical arrangement, shortly afterward the genealogical arrangement is evaluated as being obvious when Darwin writes, with respect to a tree of language, "but the proper or even only possible arrangement would still be genealogical."[5] This mutually dependent explanation of a "natural" and "genealogical" given is motivated by its subject and at the same time, constitutes the basis of Darwin's theory: the *species in a state of nature.* The method of representing the evolution of species is through their descent or genealogy: "With species in a state of nature, every naturalist has in fact brought descent into his classification; for he includes in his lowest grade, or that of a species, the two sexes."[6] Since there are "no written pedigrees," the theory of evolution must "make out community of descent by resemblances of any kind."[7] This is why Darwin regards morphology as the "very soul" of natural history.[8] Consequently, the method of evolution involves taking the morphological resemblance of the species as the point of departure for *deriving descent*. The result is a natural taxonomy or arrangement of species in a state of nature.

In contrast, a study of the evolution of culture would have to deal with subjects that have grown to a state beyond nature and whose morphology may say little or nothing at all about their meaning or function. In this respect projects of cultural evolution show that their subjects must first be isolated from the diversity and complexity of cultural phenomena—from the interplay of symbolic systems, cultural skills, bodies of knowledge, religious meanings, aesthetic practices, forms of memory and unconscious processes—in order to be constructed as units. This construction of subjects for evolutionary theory also sheds light on the "nature of the natural system." In his cultural history of evolutionary theory, Rupert Riedl thus asks:

> Have the *systematic units*—genera and families, up to orders and classes—been discovered or invented? One need only consider the following argument: If the possession of a vertebral column is assumed to be a defining feature of a group, how can it be a surprise that "vertebrates" result? These systematic units are therefore probably nothing other than inventions of our need to catalogue.[9]

Understood as a natural system, evolutionary taxonomy is indeed subject to constant improvement as newly discovered organisms are assigned a place in the system. A number of biologists are aiming to solve this problem by introducing a *PhyloCode* to replace the traditional system traceable to Linnaeus (with its hierarchy of phylum, class, order, family, genus, and species). This code is supposed to remove classificatory elements by doing away with "genus," for example, and offering definitions "solely on the basis of evolutionary principles"—as, for example, Kevin de Queiroz has stated.[10] It is no coincidence that such appeals have been voiced by the curator of a natural science museum, who is constantly confronted with questions of arrangement in the most literal sense and with taxonomy as a material culture—much like Lamarck, whose diagram of species, as will be shown, is based on problems of "distribution."

The differing degrees of acceptance that the "two cultures" (humanities and science) bring to Riedl's question of whether bio-scientific taxonomy was invented or discovered point to a methodological difference between empiric and hermeneutic processes. In contrast to Dilthey's opposition between *Erklären* and *Verstehen* as the characteristic attributes of the natural sciences (explanation) and the humanities (understanding), this difference can, in my opinion, be put to productive use: namely, to combine exact methodologies with their constant need of a reduction of complexity, on the one hand, and procedures for (re)establishing cultural contexts, on the other; i.e. to combine empirical experimental approaches and those cultural studies methods which have been developed in reference to writing, images, and concepts and can be described as reading processes.

The processes that are subject to the aforementioned reductive steps, when conceived as cultural evolution, are otherwise grasped in cultural studies under headings such as tradition and genealogy, secularization, and modernization—and also as the history of the arts and sciences. They are grasped using research frameworks, investigations, and methodological tools specific to the respective questions. Here, the epistemological interest lies not in evolution and development, but in areas such as transfer and translation, citation and reading, rewriting and conversion—i.e. in the methods by which meaning is generated or produced. Reflections on the problematic methodological effects produced by the formation of inter-temporal cultural subjects have, therefore, led to a dismissal of narratives of development, teleological models, and the history of motifs and ideas. These approaches have been substituted by studies of constellations in which attention is directed to primal scenes, i.e. arenas of emergence, by transitional figures and those non-synchronicities that, when negotiated, give rise to the concepts that have established themselves as dominant and have been canonized in cultural history. From the perspective of evolutionary theory, one could best describe this process as a kind of microscopic zooming in on those arenas in which modifications occur in the form of shifts, translations, rewritings, or ruptures.

Just when the humanities and cultural studies have bid farewell to the *grands récits,* these "grand narratives" apparently have produced an unusual boom in the natural sciences. A side effect of this renaissance is the new popularity of evolutionary theories in the field of natural history. Due to their macro-perspective, such theories seem especially well suited to showing the most diverse phenomena of "nature" in phylogenetic dimensions. And the recently observed convergence of discourses in the humanities with the knowledge cultures and language of the natural sciences has resulted in the spreading and impact of evolutionary theory in the humanities as well. A slightly different path is taken in analyses that focus on the way "systems" evolve and are based, not on a direct transfer of biological evolution to cultural evolution, but on the evolutionary concept in Luhmann's systems theory.

Present-day evolutionary theories of cultural phenomena and systems—an overview

Let us first examine the paradigm of cultural evolution in current research. If one does not use evolutionary concepts of culture in a general manner, referring to any kind of development narratives, but instead seeks applications of a stricter concept to culture, we may observe a variety of methods of "cultural evolution." They will be discussed here with respect to their relationship with biological evolution. This relationship is configured either as an *analogy*, *part* (or *highest level*) of a whole, or a *derivation*, or it is based on a *system-theoretical* link:

a) Gathered under the heading of the *evolution of culture* are studies of a human ethological, ethnological and linguistic provenance.[11] They take a *longue durée* approach to symbolic, linguistic and nonverbal communication and thus remain within the conventionalized registers that, as shown above, are probably most compatible with the methodological conditions of evolutionary theory. These studies often focus on the description of "universals"—a concept that can be effectively used to integrate cultural phenomena into the models of natural history. The *analogies* to biological evolution are most pronounced when these universals are seen as anthropological constants in the typologies of instrumental or communicative behavior. In his introductory work *Evolution–Kultur–Sprache* (1994), Klaus Bayer explains: "Over the course of generations, culture can therefore develop alongside the gene pool as a pool of acquired behavioral modes."[12] In this view the paradigm of cultural evolution appears as a compromise solution that conveys Darwin's rejection of the inheritance of acquired characteristics along with the evidence of a cultural heritage. It conceives this heritage in metaphoric terms as a set of hereditary events that occur outside genetic reproduction, though in a genetically analogous manner.
b) Quite distinct from this model is the paradigm of cultural evolution in which *culture* is understood to be a *result of an evolutionary progress*, and the further or higher development of intellectual capacities (i.e. the learning of the human species) is explained on the basis of cognitive science. In this model, "culture" arrives on the scene with *Homo sapiens*, who represent the highest level of the phylogeny of the species and the cognitive evolution of primates. Culture itself is seen as the result of "cumulative evolution." In his work *Cultural Origins of Human Cognition* (1999), Michael Tomasello, for example, proceeds on the assumption that the "cumulative cultural evolution depends on two processes, innovation and imitation (possibly supplemented by teaching), which must take place over time such that one step in the process enables the next."[13] Here, the analogy to biological evolution is replaced by the *integration* of culture into a model of descent that has been reformulated by cognitive science—culture as the highest level of phylogenetic evolution.
c) A variation of these models of cultural evolution can be found in the theories that link the *evolution of technology, language and art* more closely to the development of tools and that study their effects on the further development of "social organisms." One example can be found in André Leroi-Gourhan's magnum opus *Le geste et la parole: Technique et langage* (1964–1965). In three chapters Leroi-Gourhan outlines the history of civilization as a series of evolutionary events:

1. The emergence of technology and language, derived from the evolution of the brain and the hand;
2. Memory and rhythms, which represent the technical assistance provided for human memory by writing systems;
3. Ethnic symbols, i.e. the symbolic organization of space and time. In this model, the evolution of symbolic practices and cultural skills is not regarded as being analogous to biological evolution, but is derived from physiological development or understood as its *derivation*. As a result, this model also centers on the genesis of anthropological universals, which becomes evident in Leroi-Gourhan's thesis that the design of social space in a cross or grid pattern has universal validity across all cultures and history;[14]
4. Studies in the field of *systems theory* have the most advanced and thought-out theoretical relationship with the methodological implications of the evolutionary concept. For instance, Rüdiger Voigt's essay collection *Evolution des Rechts* (1998) focuses on "examining the empirically observable processes in the development of the law from different vantage points, asking whether a development of the legal system can be observed as this system adapts to changing environmental conditions. In this case, the evolution of the law could be understood as involving those long-term structural changes of the legal system that lead to irreversible social innovations in the sense of 'evolutionary universals.'" This question is preceded by a reflection on the problems associated with attempts to "directly adopt a concept from the natural sciences in the discourse in legal studies and the social sciences."[15] In this context, attention is called to the difficulties that arose when Humberto Maturana's model of autopoiesis was transferred to the humanities.[16] In addition to constructing a closed system referred to as "law," *Evolution des Rechts* not only pursues the problem of clear lines of development, but also focuses on the phase or stage model that forms the basis of such a view.

Systems theory and the biological theory of evolution (Riedl, Wuketits, Luhmann, Fögen)

The "systems" concept structures one of the most important interfaces between theories of evolution in the humanities and the natural sciences, yet the role and location of this concept differ significantly in the two cultures of knowledge. Whereas the introduction of the system–theoretical paradigm into biology resulted in evolution being viewed as complex set of events within the organism, the adaption of the evolutionary concept in Luhmann's systems theory led to a description of the mechanisms through which social systems change. In the first case, the emphasis is on evolution *as* events in the system, in the second, on the evolution *of* systems.

In his cultural history of biological evolutionary theory, Rupert Riedl introduces the idea of a "systemic phase" in response to the observation of an "intermeshing of the two main problems: recognition and explanation under the conditions of recursive causality." These problems resulted in part from the above-cited question of the "nature of the natural system"—that is, from the question of whether the "systematic units—genera and families, up to orders and classes—have been discovered or invented."[17] In an attempt to describe the recognition of "natural systems" as an adaptive function, the author was led to formulate a comprehensive theory of the "adaptation of the complex" which topples a number of the dogmas of classical evolutionary theory. Instead, evolution here is a recursive causality, which induces complex systems to emerge from the evolution of functional connections.[18] In this view, all of evolution—"from the laws of quanta to those of the Gothic" (Riedl)—is once again understood by means of the passed-down imagery of the family tree in which taxonomic units are presented as the "result of an ongoing process of branching out."[19] The classification or construction of units is interpreted as the effect of evolutionary events to which recognition itself belongs. The processes of branching out, classification, and comprehension are thus understood as components of one and the same system.

After Riedl made the transition "from a linear or executive causality of the evolutionary concept to its integrated or functional causality," Franz Wuketits formulated a "synthetic theory" in his account of the biological theory of evolution.[20] His synthetic theory defines evolution as a "self-organizing process" and questions the conditions of the system and the "driving forces behind evolutionary events." A series of methodological problems are attributed to this synthetic theory. According to Wuketits, the problems that form the center of the controversies around the dominance of either internal causes or external conditions in evolutionary theory—problems that include the plausibility of directedness, the irreversibility of development, the relationship between genetic information and epigenetic processes, the mediation between phenotype and genotype, as well as the debate on complexity and "essential structures"—could be transferred, with the help of the idea of *self-regulation,* to an "ensemble of interactive processes and control mechanisms that are specific to every living organism."[21] He writes, "From the system-theoretical perspective, evolution is a complex 'cycle.' It is not the result of unilaterally effective factors, but an interplay between *internal and external* conditions [of selection]—in other words, evolution is a series of complex events within the system!"[22]

In this passage, particularly regarding the issue of the driving force behind evolutionary events, a "primal scene" in evolutionary theory lives on.

This took place in the eighteenth century, when Johann Friedrich Blumenbach introduced the *nisus formativus* as an epigenetic law;[23] and Caspar Friedrich Wolff established his "theory of organic development"—the *Theoria Generationis*.[24] As a result of Wolff's theory, attention was drawn to the laws of hereditary transmission and the way organisms develop from the germ.[25] An important difference is that, in the later context of debates on evolutionary theory, the question of driving forces has been transformed to the phylogenetic level. As an alternative to the Darwinian principle of chance and mutation, today an evolutionary law has been formulated in which the rhetoric of natural systems ("growth") is fused with that of cybernetics ("self-regulation"). Thus, for Wuketits, "the emergence of new levels of complexity in evolution is rendered recognisable through use of the concept of self-organization and the methodological instruments of cybernetics and systems theory." At the same time, the production "of increasingly differentiated systems" is defined as growth: "Evolution is a growth of patterns or a *growth of complexity*" (emphasis mine).[26]

This biological systems theory of evolution has a number of points in common with Niklas Luhmann's systems theory, but there are also striking differences. Whereas the idea of self-regulation and differentiation in the sense of increasing complexity can be linked to Luhmann's theory, self-regulation on the level of "natural systems" necessarily encompasses the "interplay between internal and external mechanisms."[27] After all, metabolism is one of the most important features of organisms, which, as a result, are involved in a constant exchange with their environment. By contrast, when Niklas Luhmann, as a sociologist, talks of *system and environment* his environment has an entirely different meaning. Luhmann's two concepts are first and foremost based on a conceptual difference: the "system" is defined by its distinction from its "environment" via observation.

In Luhmann's "formulation of the conditions and results of the differentiation of evolutionary mechanisms," the *system* takes the very place that *species* occupy in the biological theory of evolution.[28] This raises the question of the system's status and genesis. But Luhmann's social system follows laws that are quite different than those of the "natural systems" of species. Regarding the laws of the system, Luhmann does not draw on the synthetic theory of evolution put forward by Riedl and Wuketits, but transfers the biologist Humberto Maturana's concept of *autopoiesis* to the social system. Maturana described the structure of living creatures as a "network of internal operations without any external effects." Accordingly, in Luhmann's "level-specific system–environment analysis" the development axis apparently plays a secondary role in his evolutionary concept. Evolution means only greater change, which takes place within the mechanisms of *variation, selection* and *stabilization*. In this respect, Luhmann's concept of evolution is linked to a process of de-temporalization, which places him in opposition to the

Darwinian theory of evolution and also renders his work problematic for studies in cultural history.[29] And when, in Luhmann's systems theory, the concept of *evolution* is substituted with that of *differentiation*, this concept comes close to a tautology: "Social evolution is a growing differentiation of the social system."[30] Luhmann's later reflections in *Die Gesellschaft der Gesellschaft* (1998) take a different turn. There focus is less on a transfer of the theory of evolution to society than on studies in the history of science that address central, yet controversial, arguments in the history of the evolutionary concept.

Consequently, in analyzing the relationship between biological systems theory and Luhmann's systems theory, one ends up with three decisive differences: (1) The relationship between the system, or species, and the environment is defined in an entirely different way. In contrast to the classical theory of evolution, which, in the case of organisms, attributes endogenous causes (mutation) to variation and regards selection as a kind of adaptive pressure that the environment exerts on the organism, systems theory claims that self-referential, autopoietic systems may be confused by disruptions of (or from) the environment, but cannot be forced to adapt;[31] (2) Luhmann's redefinition of evolution as a functional differentiation of a system tends to be associated with a de-temporalization of the viewpoint; and (3) while the biosciences examine evolution *as* events within a system, Luhmann's sociology is focused on the evolution *of* systems.

These far-reaching differences often lead to misunderstandings in the dialogue between natural scientists and humanities scholars, such as when the "evolution of culture" is examined using the paradigm of an "evolution of social systems" and associations with the two theoretical fields of reference come into play. While it is true that the same phenomena, such as art, religion or law, can be described as both social *and* cultural systems, a different perspective is brought to bear: in the case of social systems, the focus is on an analysis of functional aspects; in the case of cultural systems, on questions of the production of meaning. Beyond the fundamental doubts explained above concerning the application of the evolutionary concept to cultural phenomena (the laws governing their production and meaning differ radically from those relating to organisms) and beyond the aforementioned epistemological problems associated with constructing objects of study as *units* or systems, one has to be aware of the frictions between different concepts of "evolution" and "system" when the evolution of social systems is examined.

But what is more remarkable about the methodological questions and epistemological limits of the paradigm of the "evolution of culture" is that the transfer of Luhmann's evolutionary concept to cultural systems such as art, language, and law necessarily leads one to those historiographic problems with which cultural studies are already acquainted, such as the definition of eras, source issues, textual criticism, transmission problems, as well as the issues raised by symbolism, media, visual and textual representation, etc. This topic can be effectively studied in Marie Theres Fögen's work *Römische Rechtsgeschichten: Über Ursprung und Evolution eines sozialen Systems* (2002), which draws on system-theoretical concepts of evolution. Using the Law of the Twelve Tables, which has survived only in fragmentary form from secondary sources, Fögen provides an against-the-grain interpretation of the established, canonized history of law by concentrating on those arenas that have been excluded from it as "historically irrelevant," but which narrate the origins of the law. The book is primarily concerned with the founding myths, images, and legends (e.g. of Lucretia and Virginia) that accompanied the introduction of Roman law, but were excluded from its transmission as a legal system. As a result, the investigation is situated at a symptomatic methodological threshold between evolutionary theory and cultural studies. Whereas the differentiation of the transmitted legal system can be described from the perspective of evolutionary theory, the consideration of the sites of its origin resorts to cultural theoretical methodologies and examines the iconography and narrative techniques that are associated with the primal scenes and founding myths that precede the transmitted history. Thus, the book combines (system-theoretical) evolution and (cultural-scientific) genealogy.

On the emergence of evolutionary theory at the transition from classification to distribution (Lamarck)

In order to examine the genesis of epistemological problems of the evolutionary theory of organisms, it is necessary to take a closer look at the preconditions and emergence of evolutionary thought. To this end, the perspective will now be reversed from a discussion of the evolution of culture to a cultural scientific analysis of some constellations from the history of science that can serve to illuminate the genesis of the evolutionary concept.[32] The origins of the evolutionary concept *avant la lettre* and the explanations of evolution offered prior to Darwin are both especially instructive, since they show that current efforts to transfer evolutionary theory of organisms to cultural objects were preceded by a narrowing of the evolutionary concept to only "natural systems" in nineteenth-century Darwinism. At the same time, the Darwinian model emerged from a historio-scientific constellation characterized by a close link between theorems concerning the development of organisms and that of culture(s). In other words, before evolution became a bio-scientific concept, it had a much broader meaning—like many other specialized terms in the natural sciences.[33] The history of the evolutionary concept can be seen as a prominent case study for the manner in which paradigms in the humanities and natural sciences have become differentiated and are set in opposition to one another in the course of the nineteenth and twentieth century.

In the preliminary and early history of the evolutionary concept, the relationship between cultural and biological theories of development and evolution appears as a correspondence, whereas the transfer of images and concepts—as metaphors in the literal sense (the Greek *metaphora* = transference)—form a creative factor. Metaphors provide keywords and serve as the media of theory formation. With their help, concepts from one field of knowledge or one register of discourse are applied to another in order to facilitate a productive engagement with new questions in this second field. Thus the genesis of evolutionary theory can be studied by examining the use of rhetoric and metaphors in the texts that contributed to the formulation of the theory. In such an approach, the "evolution of evolution" is replaced by a series of transferal figures and operations: e.g. the birth of "evolution" from concepts like chain, *scala naturae* and taxonomy; the transference of a *sensibilité*, one of the keywords of the literary epoch of sentimentality to the nervous system functioning as a medium for perfectibility of species; maps and family trees as the figurative stage for phylo-genesis; competition as a metaphor that brought selection and survival into play; "negative entropy" as a metaphor that heralded the age of information theory within the genetic concept of evolution;[34] the alphabet and information as concepts initiating the interaction between the biosciences and information science, an interaction that produced the most recent period of genetic engineering, including the idea that mankind can intervene in the development of organisms.

The following discussion focuses on the historical site in which evolutionary theory emerged around 1800. This arena is of particular interest here, since it allows us to see certain correspondences between the temporalization of evolutionary theory around 1800 and its detemporalization in the present. The founder of the first comprehensive theory of evolution is widely acknowledged to be Jean-Baptiste Lamarck (1744–1829), who, as Franz Wuketits comments, took the "decisive step in the phylogenetic reinterpretation of nature's ladder."[35] Or, as Thomas Weber puts it, he was "the first to place the variability of species at the center of a comprehensive system."[36] Lamarck is generally associated with two elements of evolutionary theory, both are regarded as falsity in present mainstream scholarship when it is said that they had to be overcome in the advancement of science: his thesis of a slow or *gradual* change of species in response to environmental requirements, which Darwin replaced with an emphasis on variability under forces of selection and of the possibility of rapid change or mutation, and his thesis of the hereditary transmission of organic modifications that resulted from use. This second thesis was later labeled the "inheritance of acquired characteristics" and rejected as "Lamarckism." Whereas there is no controversy in the historiography of biology about Lamarck occupying the place of a precursor of evolution theory, in contrast, many aspects of Lamarck's theory turn out to be much more ambiguous upon a closer inspection.

The title of Lamarck's *Philosophie Zoologique* (1809) signals that it is coined by the paradigm of *histoire naturelle*, whose protagonists largely regarded themselves as philosophers and naturalists (Buffon, Maupertius, Bonnet); their aim of a comprehensive natural order found its perfect embodiment in the table of Linnaeus's system. Read as a phylogenetic reinterpretation of nature's ladder, Lamarck's zoological philosophy takes the step that is generally regarded as the temporalization of natural history or the birth of the idea of development—later expressed by the term "evolution."[37] *Philosophie Zoologique* is of greatest interest to us here because it marks an intriguing transitional point between classification and genealogy, between the table and the family tree. Until the eighteenth century, the concepts of chain and *scala naturae* coexisted well, and specifically the figure of a ladder provided a perfect condition of possibility for translating the order of species into a *succession*—that is, i.e. for a projection into the dimension of time. But for this to happen, the idea of development, which had come to the fore in the disputes between preformationism and epigenesis, had to be shifted from the individual organism to the species.[38] This shift was facilitated by the idea that different species had developed along divergent paths and were related to each other by direct descent or through resemblance or analogy.

Lamarck's great ambition was not only to broaden taxonomy, but also to introduce a more adequate order. He was convinced that his proposed system more closely approximated the actual order of nature since nature could only have produced organisms (Lamarck always speaks of *corps vivans* [living body]) *successively* starting with the simplest and only recently arriving at "les plus composées" (the most complex). The ideal goal that he hoped to achieve using his table was a "véritable disposition à donner à une distribution générale pour représenter l'ordre même de la nature" (a true arrangement to give to a general distribution for the purpose of representing the natural order itself).[39]

This rather elaborate formulation is a reference to the fact that the actual issue was representation. The desired system was a means to represent an order that approximated that of nature—that is, an order that was like that of nature itself. However, even if imitating the actual physical order of nature appears as an ideal here, there remained an awareness of the gap in the means of representation. Lamarck's subtitle, *Tableau de la distribution et classification des animaux* (*Picture of the distribution and classification of animals*) once again expresses this constellation: "suivant l'ordre le plus conformé à celui de la nature" (following the order that most approximates nature).

To give but one example of how Lamarck was received and revised in post-Darwinian years, Arnold Lang's German translation published in 1876 presents the text in the spirit of the time, translating Lamarck's text in a smoothed-out evolutionist form. For example, it renders the subtitle of Lamarck's table, quoted above, as "nach der *naturgemäßesten* Ordnung" ("according to the *most natural* order")—thus using the abovementioned Darwinian notion.[40] Also, the phrase "disposition qu'il faut donner à leur distribution générale pour rendre conforme à l'ordre même de la nature" (arrangement that is necessary to give to their general distribution in order to conform to the very order of nature) (Lamarck 1809, 269) is translated as a "naturgemässe Reihenfolge in ihrer allgemeinen Anordnung" ("natural succession in its general order").[41] Furthermore, Lang chooses *Reihen- und Stufenfolge* (sequence and graduation) to express Lamarck's proposal to replace the classes, genera, and species by a system that reflects nature's creation from the most simple up to the most complex. Finally, the six "degrees" (*degré*) mentioned in the original become, in translation, six *Stufen* (levels), to which the four "classes" of invertebrates and four classes of vertebrates are assigned. In order to assimilate Lamarck's text more closely to the myth of the evolutionary system as a natural system, the translation erased all traces of the scientific efforts to develop a method of knowledge and representation that best suits nature. Lamarck is appropriated as a forerunner of evolution by disambiguating his concepts and linguistic figures.

Nevertheless, the tension between "distribution et classification," between distribution and classification, in Lamarck's *Tableau* of animals is symptomatic of the difficult transition from Linnaeus's system to a phylogenetic theory of species. The role played by the concept of *distribution* in Lamarck's writings supports an alternative interpretation of the diagram that has been canonized in the history of evolutionary theory as "Lamarck's tree." Alexandre Métraux, for instance, recently argued that this diagram represents, not a tree, but a model that emerged from the material order of things—from Lamarck's work in the Musée d'Histoire Naturelle. In the museum's archives there he regularly had to face the difficulties of classifying individual objects and arranging them in the spatially organized order of things.[42] It is this hands-on experience with the material culture of the archives that reappears in Lamarck's theoretical concept of the "Tableau de la distribution et classification des animaux." At the same time, his diagram bears traces of a "chain"—of the metaphor of the "chaîne animale" from which the idea of a gradual order arose. Again, this primal scene illustrates that problems of classification are at the heart of reflections in evolutionary theory. In Lamarck's work, they even provide the motive for introducing a new order of living organisms.

However, the phylogenetic perspective in his theory of the development of species only really comes to bear in the third part of *Philosophie Zoologique*, where he discusses the "causes physique du sentiment" (physical causes of feeling) and claims to derive mental activity from the physiological foundations of the nervous system. In so doing, he had to move beyond the empirically researchable field of organisms. His theory links the higher development of organism to phenomena such as *sensibilité* and *sentiment*—concepts that elude both physiological access and the concept of *distribution*.

These efforts led to an evolutionary theory of living organisms that, on the basis of the nervous system, creates a hierarchy of the sensations received by living organisms: from simple sensations based on a *sensibilité physique,* to the capacity for feeling (*sentiment intérieur*), to mental activity (*l'entendement, facultés de l'intelligence*). Whereas Lamarck attributes the force of organic movement to stimulation (*excitatrice*) and irritability (*l'irritabilité*), he regards the physical causes of feelings as being rooted in a simple or more complex nervous system. But it is only the development of the brain, which he calls the "organ accessoire" of the vertebral column, as well as the formation of a center for the generation of feeling that creates the preconditions that enable mental activity, will, memory, and imagination to emerge.

Emotion is thus assigned a central place in the construction of Lamarck's phylogenesis. It is a sort of medium that conveys basic physical states together with moral capacity and thus acts as "a source of the one as well as the other"—as Lang writes it in his extensive chapter summary.[43] However, with the concept of emotion, Lamarck avails himself of a *Pathosformula* from the age of *sensibilité*—an eighteenth-century invention that could only become the medium to bridge and close the gap between physical causes and the mind, between the organic and the intelligible, in both its meaning as "la base et l'agent conservateur de la vie" and its function in combining disparate perceptions to form a *sensibilité morale.*

Cultural history as evolutionary theory *avant la lettre* (Schlegel, Herder, Schelling)

Two principles generally associated with Lamarck—the law of the strengthening or weakening of organs based on their use, and the law of the hereditary transmission of organs that have changed under the influence of the environment—are also reflected in ideas of contemporary cultural philosophy, where the concept of higher development was widely recognized. One example is Jean-Jacques Rousseau's concept of *perfectibility* in *Discours sur l'origine et les fondements de l'inégalité parmi les hommes* (1755), which portrays early peoples or cultures as the "children of mankind;" another is Johann Gottfried Herder's history of peoples, which uses the pattern of ascending forms; another still is Friedrich Schlegel's *Geschichte der alten und neuen Literatur* (1812), which recounts the history of periods (*Epochengeschichte*) as a history of generations and presents the image of a family tree in which the different literatures succeed and grow out of each other, with "modern German literature" at its capstone.[44] In Schlegel's historiography, the tree becomes the paradigm for an account of universal history that presents descent as a hierarchical gradation. In other words, succession is transformed into a series of gradations in a history whose upward movement represents the development to perfection. Whereas the ascending path of the "cultivation of the intellect" (*Geistesbildung*) is depicted as a passage through "all the eras of the world," the various national cultures embody units representing the ages of humankind. However, as levels in the development of humanity, these cultures are not accorded a development of their own. In this respect, the construction of ahistorical units—in this case national literatures—also precedes the idea of evolution.

Johann Gottfried Herder's *Ideen zur Philosophie der Geschichte der Menschheit* (1784–1791), the first volume of which was published more than two decades before Lamarck's *Philosophie Zoologique*, contained numerous elements of evolutionary theory. In addition to the metaphor of the *chain* and the model of a *succession* of cultures or peoples—a "development of beings from each other"—his study pursues the question of their directedness, or more concretely, the question of whether there was a "thread of development throughout the centuries." But his phylogeny of mankind, which was clearly inspired by the thesis of epigenesis, also examines the further development of individual peoples.[45] Like Lamarck, he undertook this examination from the perspective of environmental influences, specifically, on the basis of geographic and climatic conditions. On the other hand, the geographic order, seen as a matrix for classifying peoples (Asian, African, Greeks, Americans), represents a quasi-natural foundation whose constitution is not further reflected by Herder. Instead, this order serves for him as a subject of the descent narrative. The fact that the epistemological figures of *map* and *family tree* are superimposed on each other in Herder's account is symptomatic of the "opposition between genesis and climate" discussed by Herder—a phrase in which it is easy to recognize a leitmotif of later evolutionary theory. For example, on maps he wrote: "Thus, with the assistance of a few maps for the convenience of inspection, we should obtain a physico-geographical history of the descent and diversification of our species according to periods and climates, which at every step must afford us important results."[46] On genesis he commented: "Not only is the germ of our internal disposition genetic, as well as our body frame, but every development of this germ depends on fate, which planted us in this place or in that, and supplied us with the means by which we were formed, according to time and circumstances."[47]

In addition, one finds in Herder's work a thesis which became prominent at a much later date—one that was credited as Haeckel's Law: recapitulation hypothesis,[48] i.e. the idea that every living creature has to repeat the evolution of its species in its own ontogenesis. In this respect, Herder's concept can definitely be termed an evolutionary theory *avant la lettre*. In the philosophy of cultural history, its stage is formed by a transitional constellation that is comparable to Lamarck's project in the field of the natural history. Whereas Herder's order of peoples appears to be naturally given and his concept of development derives from the tension between map and family tree, Lamarck's natural order emerges from the interplay between the epistemological figures of *histoire naturelle* (chain/scale) and the archival method of *distribution*. At the same time, Lamarck's phylogenetic concept is based on the application of *sensibilité* as a medium between the physical and the moral. Such transfers of epigenesis to the field of culture abrogated the strict opposition between the concepts of epigenesis and evolution, as once described by Johann Friedrich Blumenbach in his *Über den Bildungstrieb* (1781).[49] The idea of an evolutionary history of mankind was created by combining the epigenesis of the individual organism with the development of peoples or cultures.

One of the first *explicit* uses of the word "evolution" in the epigenetic sense, i.e. the formation (Bildung) of the individual, can be found in F.W.J. Schelling's *System des transcendentalen Idealismus* (1800), in which Schelling speaks of the "self" ("Ich") in the midst of "the original evolution of absolute synthesis."[50] His history of the evolution of the self, which culminates in self-awareness (Selbstbewußtsein), takes its cue from the idea of perfection and from teleology. In anticipation of the scientific discussions of psycho-physical parallelism 100 years later, Schelling proceeds on the assumption of a "parallelism of nature with intelligence." According to Schelling, this parallelism can only be grasped by connecting the two fields. In contrast to Herder and Lamarck, who operate on the *threshold* between natural history and philosophy, Schelling's system aims at *collaboration* between the two fields, which he sees as counterparts to each other:

> The author's chief motive for devoting particular care to the depiction of this coherence, which is really a graduated sequence of intuitions, whereby the self raises itself to the highest power of consciousness, was the parallelism of nature with intelligence; to this he has long since been led, and to depict it completely, neither transcendental philosophy nor the philosophy of nature is adequate by itself; both sciences together are alone able to do it, though on that very account the two must forever be opposed to one another, and can never merge into one…[51]

Although there is a close proximity between such philosophical works and Lamarck's theoretical construct, it is evident that the genesis of evolutionary accounts and interpretations of both natural and cultural history stem from one and the same epistemological context. Consequently, we can say in summary that: (1) the concept of evolution arose from the interaction between paradigms in natural and cultural history; (2) the transfer of concepts and images between both fields played a productive role in this process; (3) the theory of evolution emerged from a shift in the order of things from classification to genealogy; and (4) after the development perspective became dominant, evolutionary theories still carry along the paradigm of classification as a kind of epistemological blind spot, as it were.

Lamarckism in explanations of cultural evolution (Darwin)

All the aforementioned authors who were writing around 1800 and whose reflections preceded the currently dominant bio-scientific theory of evolution share the assumption of gradual change and the evolution of species in an exchange process with the environment. They also share the conviction that acquired traits are inherited. What makes Darwin's evolutionary theory differ so radically from that of his precursors are the principles of sometimes rapid change (i.e. mutation) and of natural and sexual selection, as well as his rejection of the idea that organisms inherit acquired characteristics and modifications that result from the use of individual organs. However, the opposition between *Lamarckism* and *Darwinism*, which is commonly derived from these differences, is not tenable if one takes a closer look at the writings upon which this opposition is supposedly founded. First and foremost, as discussed, Lamarck's reflections cannot be reduced to the abovementioned thesis. But what is even more remarkable is the way Darwin treats the principles of his own evolutionary theory when he examines phenomena that exist on the border of culture and thus extend beyond those species in a "natural state" that form the basis of his natural system, as shown above.

In the last chapter of *The Descent of Man* (1871), which deals with the development of the intellectual and moral capacity of humans, Darwin's explanation largely remains within the framework of his primary argument that "man is descended from some less highly organised form." In order to lessen the divide between animals and mankind in the sense of a shared line of descent, the same conditions must potentially be ascribed to both and the distinctions between man and animal must be conceived as a series of gradual differences. As regards this issue, it is the high level of intellectual capability and moral disposition in humans that poses the greatest difficulty for Darwinian theory: "But every one who admits the principle of evolution, must see that the mental powers of the higher animals, which are the same in kind with those of man, though so different in degree, are capable of advancement....for with our domesticated animals, the mental faculties are certainly variable, and the variations are inherited."[52]

However, the developmental capability, variability, and heritability of *mental faculties* which are assumed to be shared in animals and man, are very difficult to reconcile with the fundamental principles of an evolutionary theory that categorically rules out the gradual development and hereditary transmission of acquired characteristics. This is why the idea of an ongoing refinement creeps into Darwin's thinking at this point, particularly when he writes: "The higher intellectual powers of man, such as those of ratiocination, abstraction, self-consciousness, etc., probably follow from the *continued improvement and exercise* of the other mental faculties" (my emphasis).[53]

With this idea of improvement through use, Darwin's version of evolution does not differ significantly from Lamarck's.

In terms of *moral faculty*, Darwin manages to provide an explanation that is more consistent with his theory of evolution by locating the foundation of the formation of moral characteristics in the existence of a "social instinct," a concept that also includes family bonds. Since descent is both the basis and medium of his evolutionary theory, the origin of a moral instinct that unites man and animals can be derived from the social aspect of reproduction, from these "family ties." It is precisely this aspect of Darwinian theory that offers a link to the current theories of cultural evolution that have universals as their subject. Language is always a borderline case for such theories, since *language* can be described either as a (natural) system of signs or as a system of cultural meaning whose complexity exceeds the status of universals.

Language also represents a borderline case in Darwin's work because in the case of language, the *gradual difference* between animals and man appears to become a *qualitative difference*: "Man is impelled by the same general wish to aid his fellows; but has few or no special instincts. He differs also from the lower animals in the power of expressing his desires by words, which thus become a guide to the aid required and bestowed. The motive to give aid is likewise much modified in man: it no longer consists solely of a *blind* instinctive impulse, but is much influenced by the *praise or blame of his fellows* (emphasis mine).[54]

It is not only the capability of communicating through language that makes mankind's social instinct "seeing," as it were. It is also the desire to be well remembered—that is, it is the culture of commemoration. However, inquiries that develop these thoughts further would apparently no longer fit into the framework of Darwin's evolutionary theory. After all, with the use of the term "nevertheless," all doubts are dispelled when Darwin writes soon afterward that the first foundation of a "moral sense" lies in the social instincts, including sympathy, "and these instincts *no doubt* were primarily gained, as in the case of the lower animals, through *natural selection*" (my emphasis).[55]

But even if the phrase "no doubt" eliminates doubts in this passage as to whether the law of natural selection also applies to the development of *Homo sapiens*, these doubts arise more insistently in other places—namely, in a work in which Darwin attempts to explain the expressive gestures of man and animals from the perspective of evolutionary theory. Not only is *The Expression of Emotions in Man and Animals* (1872) concerned with the *capacity* to express emotions and the question of how this capacity has emerged (a question that is answerable within the logic of his theory to the extent that the capacity can be explained physiologically). It also examines the *concrete* gestures of nonverbal body language. Regarding this type of language, Darwin states in many passages that cultural differences can be observed. For example, in the chapter on *anger* in the "natives of India," he writes: "Their gestures were very different from those of Europeans."[56] Such concrete differences disappear in his general thesis that the *expressions of emotions* that are described in his work represent a universal phenomenon that mankind shares with its ancestors. However, Darwin had much greater difficulty answering the question of whether the phenomena he examines are voluntary or involuntary, innate (i.e., inherited) or learned modes of expression.

The forms of expression that Darwin describes all belong to a nonverbal register: physical or nervous sensations that are expressed as blushing, trembling, weeping, sweating, etc., and the expressive gestures, movements or actions of individual parts of the body such as raised hands, nods, clenched fists, etc. The composition of his book is based on long since established catalogues of emotions and a chapter is devoted to each group of feelings: anger, joy, disgust, astonishment, etc. As regards the described phenomena, Darwin distinguishes the direct, non-intentional activities of the stimulated nervous system, which he attributes to "nerve-force" and which are expressed physically, from forms of expression that he regards as the effects of habit.[57] He speaks throughout of *actions, movements,* and *expressions*—concepts that can be used to grasp manifestations of the nervous system as well as individual gestures. The neuronal system takes over the place of the active subject and becomes an actor itself, particularly in Darwin's discussion of *actions* of the nerves. These expressive gestures are categorized as being involuntary and unconscious. In others words, it was the neuronal system that controlled involuntary actions—as in the work of current proponents of the idea that human actions do not follow the will.

Darwin begins his book by defining three main principles that serve as leitmotifs for his discussion. According to the principle of *habitualization,* specific emotions, desires and sensations are so closely associated with certain movements—regardless of whether they are useful or not—that they are performed habitually, even in situations in which the sensations are very weak. And even if the movement is deliberately suppressed, some of the muscles over which the will exerts the least control will still become active. A second principle is that of *antithesis* which holds that, due to these habitual movements, there exists an involuntary inclination to express oppositional impulses by oppositional movements, even if these movements are not beneficial. Finally, the principle of the *direct activity of the nervous system* states that strong stimulation of the cerebral spinal system produces effects that are involuntary and largely independent of habit.

Whereas the involuntary signs of stimulation remain within the bounds of purely physiological processes, it is particularly interesting for the question of cultural evolution to study how Darwin interprets and explains habitual expressive gestures. Based on the assumption that those gestures that can be attributed to habit are not learned by the individual, but are already developed in small children, Darwin writes that they must be seen as "innate or inherited."[58] Thus, these phenomena touch upon problems associated with the history of species, or the Descent of Man—problems that he endeavors to resolve in his "Concluding Remarks and Summary," the final chapter of *The Expression of Emotions*: When did these expressions develop? What role is played by use and the will? In what way were these expressions transmitted to different generations and "races"? For Darwin, an explanation of habitual expressive gestures that does seem to go beyond the framework of his own evolutionary theory is provided by the thesis that, in the course of evolution, the remains or rudiments of originally useful, deliberate, or consciously performed actions have become involuntary, habitual, and hereditary. The idea that gestures evolve from being habitual to being hereditary—which makes them part of the process of descent—is usually seen as a feature of Lamarckism or associated with the hereditary transmission of acquired characteristics. But it is precisely this explanation that then raises the question of how one must imagine such a development.

A closer examination of Darwin's explanations reveals that, on a rhetorical level, he operates with extremely vague phrases, e.g. "at first," "soon," and "at last."[59] What is even more remarkable is that *culture* makes a brief appearance as an argument but then disappears from the text altogether. The examples Darwin cites to expand on his rudiments thesis include the obliquity of the eyebrows of a person suffering grief or fear, and the contraction of the muscles around the eyes when a person weeps, which Darwin attributes to an original physiological protective function. This is the place where culture comes into play:

> This action, in the course of many generations, has become firmly fixed and inherited: but when, *with advancing years and culture*, the habit of screaming is partially repressed, the muscles round the eyes still tend to contract, whenever even slight distress is felt. …Slight movements, such as these just described, or the scarcely perceptible drawing down of the corner of the mouth, are the last *remnants or rudiments* of strongly marked and *intelligible* movements. They are as *full of significance* to us in regard to expression, as are ordinary rudiments to the naturalist in the classification and genealogy of organic beings (emphasis mine).[60]

In other words, Darwin assumes that the process of habitualization was part of an early period of evolution, and the advancement of culture led to expressions of emotion being suppressed. These expressions did not disappear completely but became involuntary and now take the form of the rudiments of this early history of mankind. It is thus the history of culture that transformed formerly voluntary movements into involuntary and unconscious ones.

This interpretation could contribute much to a theory of the unconscious. And, indeed, the dialectic between censorship, or the repression of instincts, and unconscious expressions was central to Freudian psychoanalysis. But it confronts Darwinian theory with a number of contradictions. According to this view, either the will preceded involuntary actions in evolution, or the processes defined as "evolution" were concluded with the appearance of culture and its effects, meaning these effects would have to be examined using other methods. One example of such a method can be found in Walter Benjamin's Baudelaire book, in which Benjamin describes an entire repertoire of involuntary tactile actions that emerged in modern times as a result of the development of apparatuses, media, and instruments.[61] But Darwin takes his argument in an entirely different direction. In the last sentence of the cited passage, he steers the question of the meaning of intelligible movements into the context of the descent theory. His statement that, with regard to expression, these movements are "full of significance" does not introduce—as one might initially expect—a study of their semantics. Rather, it is employed as an argument establishing an analogy to the taxonomy of organisms: rudiments are important because they can be seen as indicators of the position of species within evolution. In the end, Darwin's *Expressions of Emotions* is not about explaining these expressive gestures. It is about using them to prove the arguments of his descent theory.

Another approach to resolving these difficulties was taken by the biologist August Weismann—initially a supporter of both the recapitulation thesis and the idea of the inheritance of acquired characteristics. Weismann's attempt to explain the intellectual and cultural development of mankind led him to revise his earlier position and narrow the scope of the inheritance theory by making a distinction between *hereditary transmission* and *tradition.* In his view, tradition is the capacity to pass on "intellectual ability to descendants so that they gain the advantage of being able to place themselves on the rung achieved by earlier generations and rise from this rung to a higher one."[62] When progress is linked to the succession of generations, it becomes clear that the concept of generation often serves as a representative of culture in the biological inheritance theory.[63]

Let us return to Darwin's text. After culture (or cultural history) makes a brief entrance in his book, he touches upon a number of gestures that, although appearing artificial or conventional, are in fact innate, and he evaluates, as exceptions in this context, individually learned gestures such as praying hands. Then his argument returns to the question of descent. What follows is an internal discussion of issues in evolutionary theory, including the universality of expressive gestures, particularly with respect to the problem of these gestures being found in all races. Darwin assumes that, in the course of evolution, these races only became differentiated by natural selection. His goal here is thus to reconcile a universal attribute with the differentiation of species by selection. To this end, he takes the assumed universality of nonverbal expressions as proof that the different races descended from the same progenitors, who already had a human character.[64] On the basis of this thesis, Darwin is able to trace the emergence of individual expressions of emotion to different steps within the evolutionary history by tying the respective movements to the necessary physical conditions in the nervous and muscle system. For example, he writes: "Therefore weeping probably came on rather late in the line of our descent."[65]

Seen in this light, Darwin's *Expression of Emotions* presents an evolutionary theory of universals in which individual expressive gestures are regarded as stages in the line of human descent. The *meaning* of these expressive gestures is thus split into two different aspects: while they are understood as *indicators* of a line of descent, their *semantics* disappear from evolutionary theory along with the concept of culture. This primal scene of the "evolution of culture" in the history of science thus makes clear how genuine cultural questions are cancelled out on the one hand side in the paradigm of heredity—and on the other hand are integrated into the paradigm of the inheritance of acquired characteristics. This latter paradigm must therefore be seen as an advance post of evolutionary theory, one in which explosive epistemological problems regarding the interplay between nature and culture are negotiated.

Translated from the German by Adam Blauhut.

Notes

1. Richard Dawkins, *The Selfish Gene* (1976) (Oxford: Oxford University Press, 1989), 192.
2. Wilson's goal is explaining "mental phenomena" in physical and neurological terms (Edward O. Wilson, *Consilience: The Unity of Knowledge* (New York: Alfred A. Knopf, Inc., 1998), 182ff. His theory ends more quickly in metaphysics than does Dawkins' since the concepts and methods it works with are among those mythemes that, like the aesthetic power of geniuses, the talk of the "creative mind," or the interpretation of literature as an archive of archetypes, belong on the junk heap of abandoned methods in the humanities and cultural studies.
3. Wolfgang Wieland, "Evolution," in *Geschichtliche Grundbegriffe*, eds Otto Brunner, Werner Conze and Reinhart Koselleck (Stuttgart: E. Klett, 1975), II, 5, 201.
4. Charles Darwin, *On the Origin of Species by Means of Natural Selection* (1859) in *The Essential Darwin*, ed. Julian Huxley (Mineola, N.Y: Dover Publications, 2006), 5,264.
5. Ibid., 265.
6. Ibid., 266.
7. Ibid., 267.
8. Ibid., 272.
9. Rupert Riedl, *Riedls Kulturgeschichte der Evolutionstheorie: Die Helden, ihre Irrungen und Einsichten* (Heidelberg: Springer Publishing, 2003), 171. Unless otherwise indicated, all translations from the German are by Adam Blauhut.
10. "Schafft Linné ab! Der Zoologe Kevin de Queiroz will eine neues Ordnungssystem der Biologie einführen," *Frankfurter Allgemeine Sonntagszeitung*, May 29, 2005, 57. Kevin de Queiroz is curator of the National Museum of Natural History in Washington, DC.
11. See Udo Figge, ed., *Die Kultur und ihre Evolution in humanethologischer und semiotischer Perspektive* (Bochum: N. Brockmeyer Inc., 1995).
12. Klaus Bayer, *Evolution—Kultur—Sprache* (Bochum: N. Brockmeyer Inc., 1994), 44.
13. Michael Tomasello, *Cultural Origins of Human Cognition* (Cambridge, MA: Harvard University Press, 2001), 39.
14. Concerning the problems of this type of supra-historical description of urban topography, please refer to my essay, "Text und Topographie der Stadt: Symbole, religiöse Rituale und

Kulturtechniken in der europäischen Stadtgeschichte" in Sigrid Weigel, *Literatur als Voraussetzung der Kulturgeschichte* (Munich: Fink, 2004), 48–284.

15. Rüdiger Voigt, *Evolution des Rechts* (Baden-Baden: Nomos Publishers, 1998), 7.
16. These problems can be studied in the essay collection *Der Diskurs des Radikalen Konstruktivismus*, ed. S. Schmidt (Frankfurt am Main: Suhrkamp, 1987).
17. Riedl, *Riedls Kulturgeschichte*, 170ff.
18. Ibid., 210.
19. See Sigrid Weigel, *Genea-Logik: Generation, Tradition und Evolution zwischen Kultur-und Naturwissenschaften* (Munich: Fink, 2006), Chapter 1, "Genealogy: Iconography and Rhetorics of an Epistemological Figure."
20. See Franz Wuketits, *Grundriss der Evolutionstheorie* (Darmstadt: Wissenschaftliche Buchgesellschaft, 1989), 139 on Riedl.
21. Ibid., 135.
22. Ibid., 138.
23. Johann F. Blumenbach, *Űber den Bildungstrieb* (Göttingen: Johann Christian Dietrich, 1791), 31.
24. Caspar F. Wolff, *Theoria Generationis* (1759), trans. and ed. Paul Samassa (Leipzig: Wilhelm Engelmann, 1896), 5.
25. See Weigel, *Genea-Logik*, Chapter 5, "The Forgotten History of the Concept of Generation."
26. Wuketits, *Grundriss der Evolutionstheorie*, 141.
27. Ibid.
28. Niklas Luhmann, "Evolution und Geschichte," in *Soziologische Aufklärung 2: Aufsätze zur Theorie der Gesellschaft* (Opladen: Westdeutscher Verlag, 1991), 152.
29. This observation corresponds with David Wellbery's argument that systems theory encounters its limits when dealing with genealogical questions (Wellbery, "Die Ausblendung der Genese: Grenzen der systemtheoretischen Reform der Kulturwissenschaften," in *Widerstände der Systemtheorie: Kulturtheoretische Analysen zum Werk von Niklas Luhmann*, eds Albrecht Koschorke and Cornelia Vismann (Berlin: Akademie Verlag, 1999).
30. Niklas Luhmann, "Systemtheorie, Evolutionstheorie und Kommunikationstheorie," in *Soziologische Aufklärung 2: Aufsätze zur Theorie der Gesellschaft* (Opladen: Westdeutscher Verlag, 1991), 197.
31. See Claudio Baraldi, Giancarlo Corsi, and Elena Esposito, *GLU: Glossar zu Niklas Luhmanns Theorie der sozialen Systeme* (Frankfurt am Main: Suhrkamp, 1997), 55ff.
32. An excellent discussion of the history of biological evolutionary theories can be found in Wolfgang Lefevre, *Die Entstehung der biologischen Evolutionstheorie* (Frankfurt am Main: Ullstein, 1984).
33. See Weigel, *Genea-Logik*, Chapter 5.
34. Ibid., Chapter 10, "The Text of Genetics between Metaphor and Algorithm."
35. Wuketits, *Grundriss der Evolutionstheorie*, 21.
36. Thomas Weber, *Darwin und die neuen Biowissenschaften: Eine Einführung* (Cologne: DuMont, 2005), 82.
37. It is even more surprising that Lamarck plays no role in Arthur O. Lovejoy's *The Great Chain of Being* (1936), a classic of the temporalization theorem.
38. See Weigel, *Genea-Logik*, Chapter 5.
39. Jean-Baptiste de Monet Lamarck, *Philosophie Zoologique, ou Exposition des considérations relatives à l'histoire naturelle des animaux*, 2 vols (Paris: Dentu Publishing, 1808), 1, 271ff.
40. Jean-Baptiste Lamarck, *Zoologische Philosophie*, trans. Arnold Lang (Jena: Hermann Dabis, 1876), 145.
41. Ibid., 140.

42. Alexandre Métraux, "Von der Körperanatomie zur Textarchitektur: Űber die Verteilung der Lebewesen in der Biologie um 1800" (unpublished manuscript at the Zentrum für Litertur- und Kulturforschung Berlin, 2001).

43. Lamarck, *Zoologische Philosophie*, Lang trans., 509.

44. Weigel, *Genea-Logik*, Chapter 8, "Before and Beyond the Concept of National Literature."

45. Ibid., Chapter 5.

46. Johann Gottfried Herder, *Outlines of a Philosophy of the History of Man*, trans. T. Churchill (London: J. Johnson Publishing Company, 1803), 332.

47. Ibid., 406.

48. August Weismann also represented this position, see also Löther 1990.

49. Weigel, *Genea-Logik*, Chapter 5.

50. Friedrich Wilhelm Joseph von Schelling, "System des transcendentalen Idealismus" (1800) in *Ausgewählte Schriften*, ed. Manfred Frank (Frankfurt am Main Suhrkamp., 1985), 1I, 465.

51. Ibid., 399.

52. Charles Darwin, *The Descent of Man, and Selection in Relation to Sex* (1871), with an introduction by James Moore and Adrian Desmond (London: Penguin Group, 2004), 679.

53. Ibid., 680.

54. Ibid., 681.

55. Ibid., 682.

56. Darwin, *The Expression of the Emotions in Man and Animals* (1872), with a preface by Konrad Lorenz (Chicago: University of Chicago Press, 1965), 347.

57. Ibid., 348.

58. Ibid., 350.

59. Ibid., 352ff.

60. Ibid., 350.

61. Walter Benjamin, *Gesammelte Schriften*, eds Rolf Tiedemann and Hermann Schweppenhäuser (Frankfurt: Suhrkamp, 1980), 1, 2. Also see my work on the relationship between Benjamin's media and cultural theory (Sigrid Weigel, "Detail, photographische und kinematographische Bilder: Zur Bedeutung der Mediengeschichte für Benjamins Kulturtheorie," in *Walter Benjamin: Die Kreatur, das Heilige, die Bilder*, Frankfurt am Main: Fischer Taschenbuch Verlag, 2008).

62. August Weismann, *Aufsätze Über Vererbung und verwandte Fragen* (Jena, Germany: G. Fischer, 1892), 612.

63. See Sigrid Weigel, "Zur Dialektik von Geschlecht und Generation um 1800. Stifters Narrenburg als Schauplatz von Umbrüchen im genealogischen Denken," in Sigrid Weigel, Ohad Parnes, Ulrike Vedder, and Stefan Willer, *Generation: Zur Genealogie des Konzepts—Konzepte von Genealogie* (Munich: Fink, 2005), 109–24.

64. Darwin, *The Expression of the Emotions*, 360.

65. Ibid., 361.

Communicating vessels: On the development of a theory of representation in Darwin and Warburg

Sabine Flach

"At last a book that helps me," Aby Warburg noted in the margin of his copy of Charles Darwin's *The Expression of the Emotions in Man and Animals* in 1888.[1] Nearly 40 years later, on December 29, 1927, he elaborated:

> To conceive of human expression in visual art as a template for a practical and active life—be it as a religious cult or as the drama of court ceremonial or indeed the theatre—is an idea I have wanted to develop for a long time. It first came to me through my encounter with two books that I read during my time in Florence in 1888, entirely independently of one another and without any hope of them ever directly feeding into my thoughts on art history. These books were Darwin's *On the Expression of Mind* and Piderit's *Mimik und Physiognomik*. The fact that the general facial expression is a reflexively repeated expression reacting to a purely intellectual stimulus as if remembering a more manifest state of stimulation is, of course, the main topic of Darwin's book; an example: if one dislikes somebody, one pulls one's mouth as if one were tasting something sour. This perfectly simple maxim of the study of expression—stimulus memory following the law of maximum exertion in facial expression movements—led, over the decades, to two unexpected expansions: to the expansion of the field of observation in purely geo-(VIII) graphical and thus national terms. In this context, it was, of course, necessary to understand Darwin's theory of the memory of manifest states of stimulation not as a simple, but rather as a p o l a r p r o c e s s, in the sense that stylistic changes were not simple, dynamic reactions, but rather a backlash against pressures which I had initially perceived as "foreign influences" on Italian art.[2]

Though engaged with Darwin's ideas on expression and memory, there appears to be nothing specific about evolutionary theory in Warburg's passage. Such evolutionism was, of course, in tune with the spirit of the age and did influence both Warburg and many of his contemporaries. It is my initial hypothesis that by homogenizing and appropriating Warburg into a tradition within art history which applies evolutionary theory as a blanket influence upon art and cultural history or discusses figures like Warburg within an intellectual climate that was dominated by the life sciences, one would fail to take into consideration the special relationship that characterizes both the *latent* and the *explicit* correspondences between Warburg and Darwin.[3]

From my perspective, the case of Warburg and Darwin is less about influence than it is about correspondences. The image of *communicating vessels* comes to mind. "Communicating vessels" by analogy operate not on the basis of direct transfer or influence, but rather in the more uncertain field of correspondences. By their very nature, correspondences are based on shifts of knowledge and meaning, on allusions and intimations, on indirect and hidden correlations; they can involve interfaces between individual thought process and conclusions and the tremendous effort to make intelligible one's own knowledge and to inscribe it in a different field or, rather, to alert those working in that other field to what their watchful eye may have missed. Rather than trace an unmediated liaison between Warburg and Darwin, I intend to uncover and highlight certain correspondences. The questions I posit are: How are the "communicating vessels" in question organized, or rather, what kinds of shifts are we dealing with and, specifically, what interested Warburg about Darwin's *The Expression of the Emotions in Man and Animals*?

My guiding perspective in this enquiry is a fundamentally dual one: I am interested in a Warburg-based re-reading of Darwin as cultural history, which will, in turn, form the basis of an interpretation of Warburg's methodology. Such a perspective, however, may seem to effect the analytic arsenal at one's disposal insofar as it poses the question as to how one should ponder Warburg's method against the backdrop of his own repeated and enthusiastic references to Darwin throughout his oeuvre and to the scientist's importance to his theory of expressive gestures—or, as Warburg puts it in the introduction to the "Mnemosyne-Atlas," his theory of a "kunstgestaltende Gebärdensprache". What I am interested in above all is why a work rooted in evolutionary biology could become so relevant to Warburg, a scholar, after all, of culture and the arts. Warburg's interest in Darwin, I will argue, is not rooted in a reading of *Expression* first and foremost as genealogical evolutionism. In so doing, I will also suggest a revision to standard readings of Darwin's imagery and will propose an interpretation that conceives of his methods as a productive inter-relationship of biology and other discourses, specifically discourses of art and art theory. By re-reading *Expression* through a Warburgian prism, I will suggest that representation in Darwin's treatise then can be understood as an epistemological journey between two worlds—that is, as cultural studies *avant la lettre*. In changing perspectives on Darwin's book from science per se to culture, *Expression* can be read as a testament of Darwin's earnest endeavor to develop a *theory of the corporeal image act of display* as a scenario as it relates to expressive gestures.[4] Such an approach can be related here to the sub-phenomena of "showing," such as affect, traces, gestures, and expressions, all of which are corporeal phenomena, the subject matter and significance of which are revealed through image-generated meaning. The argument followed in this way is thus always an *iconic* one, in which visual representation always interlocks with the corporeal image act of showing;[5] this act, on the one hand, denotes the generation of material images (a drawing, say, or a photograph), and, on the other hand, transforms the body itself into an image.

Viewed from this perspective, Darwin—read alongside and through the prism of Warburg, I argue—is concerned with uncovering the conditions of symbolic action. I would suggest that through explicit reference to cultural phenomena, Darwin's treatise becomes in its turn, a bio-scientific theory of expression and representation in relation to body images, the central insight of which is intrinsic oscillation: oscillation between nature and culture, imagination and imago, factuality and fictionality, and, where our knowledge of the link between expression and emotion is concerned, oscillation in terms of gestures and details as the interplay of *habitus, habitat,* and *heredity*. The body image in question thus encompasses and unites mental pictures and representations—notably in French both are denoted by the single term *représentation*. We can also invoke the neurosciences, which distinguish between *external* and *internal* representation, not least in order to explain the generation and meaning of emotions. Upon closer inspection, it makes relatively little sense to formulate a categorical opposition between internal and external images. Instead, we need to conceive of—in the words of Serge Tisseron—"the psychological and the material image as two facets of the same process."[6]

We will now contemplate a tableau on which a variety of lines of enquiry intertwine. These are, as mentioned, characterized by the interdependence of a pair of conceptual aspects. At issue is the relationship between Warburg and Darwin, but also the interactions of nature and culture, drawing and photography, imagination and imago, or rather factuality and fictionality, and body and image. This fundamentally doubled codification, will allow us to study a representation of emotions characterized by the intrinsic and inevitable interwovenness of internal and external representation. In order to arrive at a workable methodological framework, I will concentrate on the relationship between body and image, setting aside literary language and the intricate deployment of metaphor, both of which can be found in abundance in both Darwin and Warburg.

Interestingly, both Warburg and Darwin often focus their attention on the whole body, or rather, on the body in motion, in order to study the relationship between expression and emotion (Figure 6.1). The body in

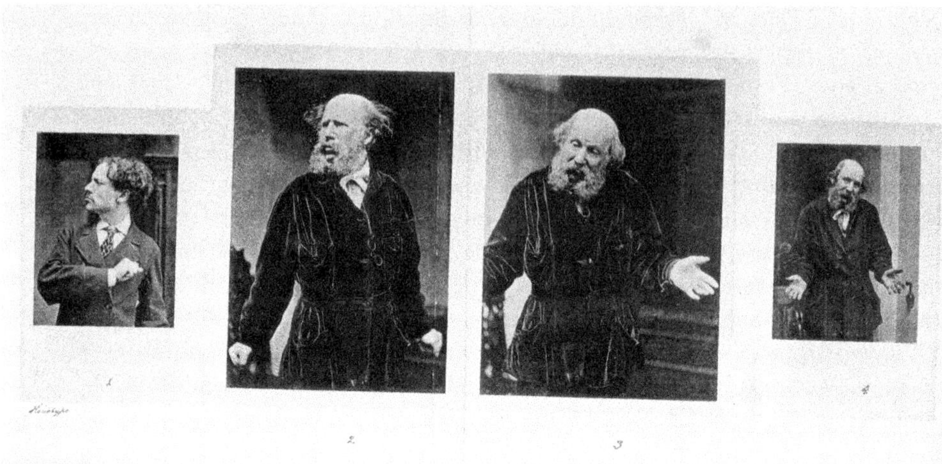

6.1 Plate VI of Charles Darwin, *The Expression of the Emotions in Man and Animals*, 1872. Indignation and Helplessness. Heliotype print. Oscar Rejlander

these cases marks the site on which images of the body, as well as images generated by the body, can be negotiated. At the same time, the expressive body always functions as the image–act for these reproductions. Boundaries between internal representation, as the body's own process of image generation, and external representation are, as the French term *représentation* suggests, permeable and anything but mutually exclusive. And if, according to Warburg, our "social memory" does indeed articulate itself in the form of expressive corporeal images, the ability to decode such images becomes a fundamental prerequisite for the cultural production of knowledge. The image can thus no longer be regarded as a "mere" representation, but rather needs to be contemplated in terms of its perceived reality, its cultural features.

In order to find a methodological anchor for a concept of the image in which the relationship between expression and emotion relates to the internal and external images of the body and to drawing and photography, I revisited two essays by Michel Foucault. In "La peinture photogénique," first published in February 1975 in the catalogue that accompanied the exhibition *Le désir est partout*, as well as in the 1982 essay "La pensée, l'émotion," Foucault developed a concept of the image that avoids turning pictures into static objects of knowledge.[7] Instead, and notwithstanding the numerous intellectual twists that occur in relation to the image from one essay to the other and result in significant conceptual shifts and divergences, Foucault posits for the image what he refers to as a fundamental *intelligibility*. The most eminent feature of this intelligibility is its event-character and thus its irrepressible suggestion of potential movement or changeability. In this perspective, images are no longer "captured," but rather occur or, in Foucault's words, "happen," though they can also wither and fade. This is a notion of the image in process, which is characterized by the slide and flow of that which is to be represented and the way it is grasped by the viewer in terms of its immediate history. In Darwin's treatise, expression consolidates itself at an extreme point as in Figure 6.1, but in capturing heightened emotions the image implicates corporeal energy that has been activated and will shortly change again.[8] In this way, the image here becomes a relay, a hub, in which events can multiply, augment one another, and be evaluated in the process.[9] Such explicit energetics of the image, which allow the image to "happen" in the mind of the viewer, suggest Aby Warburg's "Physik des Engramms" (or, indeed, of Walter Benjamin's dialectical image); the image here becomes a transformer, a converter. In "La pensée, l'émotion," the transformer makes way for the intimate circulation of a "pensée–émotion:" an intellectual or "thought–emotion." He brings up both "sensation"—which, in French, also denotes an event or an experience—and "expériences." In this essay, Foucault is no longer concerned with concrete examples of works of art; rather, the image as such discloses its own epistemes: what is generated is a form of knowledge that only reveals itself in the iconic. Thus, to apply Foucault it is no longer a question of "illustrating" a theory or of uncovering a text that underlies or lies hidden behind the image. Rather, the content conveyed by the image can only, and exclusively, be worked out from and through the image itself. The results of such work, however, are only temporary, since these images, as *representations*, are characterized by the fact that they allow events and experience to escape again, that they are ultimately incoherent, and that they elude, according to Foucault, communicability; that is, linguistically reconstructive representation.

It is precisely this iconic potential that underscores the relevance of image–act theory as a scenario. An image–act can be described as the opening up of a space of possibilities. What becomes "showable" is the potential of the undetermined in which the body can, nevertheless, appear to momentarily define itself.[10] Image–act theory operates with transitions, indeterminacies, ambiguities, and energetics and thus generates its own eidetic verisimilitudes. Against the backdrop of this methodological approach, I will now concentrate on three specific aspects taken from the rich and varied correspondences between Warburg and Darwin. Focusing on the relationship between expression and emotion, with a view to formulating a representational theory of the image–act, I will address the following areas: (1) the significance of contingency and variability through movement; (2) "energetic transformation," i.e. imagination and imago, as well as factuality and fictionality; and (3) gestures as habitus in habitat with heredity.

Darwin develops a new intellectual sphere: notions of variability and contingency

When Darwin first developed his ideas, he would have found it difficult to garner much support for his specific approach. The intellectual climate of the period was still dominated by natural theology, according to which all organisms were designed by god, who had adapted their form perfectly and permanently to their specific habitats. The dominant perspective on nature was thus a teleological one, which saw the products and phenomena of nature as part of a large and single continuum. Darwin's break with natural theology, and with the tradition of thinking in linear genealogies, could hardly have been more radical or pronounced. Theories of the mind and expression were largely tied to natural theology as well.

Yet philosophers of the English Enlightenment—Locke, Reid, or Berkeley, to name but a few—had shown a keen interest in the complex processes of construction by which the brain models and depicts reality without relying on religion or morals and the relationship a thus *embodied* reality then has with these processes.[11] What was at issue was the question as to how perception related, on the one hand, to a constantly shifting environment and, on the other, to past (and thus memorable) sensory and emotional experiences, though it should be noted that Locke established a tradition whereby memory was based on the experiences within an individual's lifetime and not on evolutionary history. Later, in the nineteenth century, aesthetic theorists continued to consider the problem of the interaction of neuronal and perceptual processes, feeling or sensation and the human capacity for expression, from a wide variety of angles. Robert Vischer, in his 1873 treatise "On the Optical Sense of Form," distinguishes, for example, between sensation and feeling (a distinction, by the way, that corresponds closely to that between emotions and conscious feelings commonly drawn in contemporary neuroscience). According to Vischer, sensations—located at a more basic level in the perceptual process—arise "intuitively" and precede conscious feeling. Conscious feeling, in turn, precedes our capacity for empathy—a capacity that manifests itself in and through an object's (outward) form. Each of these states is, according to Vischer, linked, on the one hand, to what he calls "motor nerve function" and to the perceived shape of an object in relation to one's corporeal form.[12]

The images employed by Darwin play an important part in the revolution his work set in motion—notably, they do not function merely as secondary visual aids to an independently developed theory; on the contrary it is *with* and *through* these images that Darwin is able to project an intellectual horizon that leaves room for variability and contingency; that is to say, for the transitory effects of emotional expression in which one can observe the workings of evolution. The intelligibility of the wood engravings and photographs

Darwin used assumed center stage; the task was to acknowledge a state of motion as constitutive of an exploration of expressive movement. Darwin's system stood in opposition to rigid typological systematization—rather, I argue, it was based on blurred boundaries and proliferating variations of facial expressions, gestures, and other physical features.[13]

In order to gain a deeper understanding of the function of "expression" in relation to "emotion," Darwin not only closely observed both animals and his own children, but also corresponded with a large number of physiologists, among them Frans Cornelis Donders of Utrecht, whom Darwin also commissioned to conduct a study on lachrymal secretion. He carefully consulted numerous pertinent works—for example, the writings of Duchenne de Boulogne and Charles Bell. On November 21, 1826, Charles Darwin attended an event at which the gathered participants sought to refute certain hypotheses put forward in Charles Bell's 1806 book *The Anatomy and Philosophy of Expression as Connected with the Fine Arts*. Bell had argued from a perspective of natural theology that a thorough understanding of the anatomy of the emotions could help artists in the creation of their works; scientific knowledge also formed the basis for artistic production. He stated, "Anatomy [is] the examination of the structure by which the mind expresses emotion and through which the emotions are controlled and modified; it introduces us to the knowledge of the relations and mutual influences which exist between the mind and the body. To the painter, therefore, the study is necessarily one of great importance; it does not teach him to use his pencil, but rather it teaches him to observe nature, to see forms in their minute varieties [and] to catch expressions so evanescent that they must escape him, did he know their sources."[14] In his book, Bell had tried to warn a new generation of artists against the mistake of devoting too much time to the study of antiquity at the expense of a study of nature when tackling the problem of representing expressive gestures. In Bell's advocacy of the study of nature he denied animals the full range of expressions that humans are capable of, largely, though not exclusively, on the grounds that god has given to man, but not to animals, the capacity of speech and the mobility of certain muscles to communicate with other humans. Charles Bell was thus convinced that an understanding of the physiology of the emotions could help artists to "translate" an emotional state into a work of art; that is, to make emotions *communicable*, and thus to find themselves supported by the work of science in a useful and productive way.

From his earliest writings onwards, Darwin disagreed with certain central aspects of Bell's views. His notes, especially after the *Beagle* voyage, include records of his observations of expressive states of both men and animals that would, according to Bell's analysis, have been impossible. In a letter to Alfred Wallace in March of 1867, Darwin noted, "I want, anyhow, to upset Sir C. Bell's view [...] that certain muscles have been given to man solely that he may reveal to other men his feelings."[15]

Darwin's image archive is not concerned with a divinely inspired genealogy of expressions, and in his physiological system, his explication is preceded by the visualization of a given gesture. While Darwin believes that these gestures are, at least in part, responses to nerve stimulation, they also—and crucially—have their roots in habit or even occur at random.[16] Nevertheless, every expressive gesture comes with its own history for Darwin; this history is based in evolutionary biology or from individual habit, and by virtue of its locus in internal structures, potentially variable. It is at this point that the concept of the *engram* can be invoked in Darwin's theory. In scientific terms, the engram denotes the trace left by a particular stimulation or experience on the central nervous system; this trace then facilitates the impression's reproduction at a later point in time. In cultural studies, meanwhile, the term engram refers to a memory image. A link between biologism and cultural enquiry can lead us back to Warburg's reflections on the "engrams of passionate experience" as memory's treasure-trove and allows us to establish a model in which the body image is based in memory of the past, experience of the present, and the performance of the moment.[17] At the same time, such a notion of the body image highlights the fact that emotional processes arise from the complex interplay of body and mind, and that these processes, in turn, permanently overlap with ideas and images about emotions and the appearance of an expression already stored in the brain (as argued by eighteenth-century associationists like Locke). To cite Thomas De Quincey's 1845 essay, "The Palimpsest of the Human Brain," such processes can be described as "endless strata...not as a succession, but as parts of a coexistence."[18] For Darwin, for an image theory of expression to be realized in the first place, the viewer / reader would need to be able to take into realistic consideration contingencies, as well as the fleeting and barely visible. What takes place here is a concentration on a conceptual space opening up that would previously have been inconceivable precisely because it suggests that which is elusive. *Expression* thus plays on the boundaries of the visible in so far as that which is viewed as a concrete and measurable state of affairs is forced into the background in favour of a representation of fleetingness that harbours the *possibility* of a transforming visual manifestation.

Darwin developed this principle of contingency through images, since images bring the distinctive properties of life into tight focus: Darwin is particularly interested in the shifting aspects of expression, with the body passing through a sort of "Morse code" of body movements that translates into kinetic energy. By emphasizing expressions as subliminally effective graphemes, Darwin departs markedly from static symbols and takes, in my view, a crucial step—namely, a step away from traditional naturalistic representation, as generally understood by artists at that time, towards the "abstract" representation of physical expressions *in motion*, as well as towards the representation of the inherently moving image. Representation thus becomes a force field in a Warburgian sense, momentarily captured and arrested, that appears nevertheless to be able to escape their arrested state at any given moment.[19] In this way, naturalistic appearance from a traditional perspective is shattered and its reconceptualization can then serve as building blocks for a new theory of representation.

"Energetic transformation" or, the relationship between nature and culture, factuality and fictionality

The above perspective situates Darwin at the threshold where nature and culture are not conceived in opposition to one another, but are rather seen as fundamentally entwined. I would like to cite Darwin's use of images as one of the most prominent and palpable examples of this kind of entanglement. It is the *modifications, variations,* and *inscriptions* that such images are subject to in this scheme of things that are invested with particular meaning. These modifications and variations are what I will be addressing in the course of the following pages.

Figure 4 of Darwin's *Expression* features the image of a small dog watching a cat on a table, though only the dog is visible. The original of the image reproduced in *Expression* is not, as one might imagine, a drawing, but rather a photograph taken by Oscar Rejlander (who appears as the man with the grey hair in Figure 6.1). For its inclusion in *Expression,* Darwin, who, as we know, did use photographs in a variety of contexts, had commissioned a wood engraving of the image, which was then reprinted in his book. The significance of this deliberate change in media should not be underestimated. In Darwin's day, photographs were, after all, regarded as documents with the highest claim to truthfulness, a claim that was legitimized by the supposed absence of any outside intervention. To the extent to which technical recording devices—with their inherent conformity to the ideal of the machine—were associated with neutrality, transparency, life-likeness and thus with the highest cogency of proof, these qualities were perceived to be lacking from drawings, which were instead associated with the arbitrariness that arises from the inherent subjectivity of the artist and observer. As such, the photograph marks the pinnacle in the contest between drawing (including the process behind wood engraving) and technical recording devices.

Darwin's chosen method of deriving an image (in his text) from the image (wood engraving) of an image (photography) introduces small, but nevertheless significant and marked modifications. If an image, according to its representational status, is required to be both like and unlike that which it seeks to represent, i.e. if, in order to be and remain an image, the image cannot simply reproduce all the elements of its object of representation, *difference,* as an element of form, must be regarded as constitutive. This form of image-repetition is thus *not* the recurrence of the identical, of the one-and-the-same.[20] The deliberate intervention in the image on the part of a draughtsman marks a process of inscription that makes visible a methodological knowledge as knowledge about the consequences of implied action. What this process thus ultimately makes visible are the different epistemic styles of the image itself, which, in its turn, is tied to experiential knowledge and based on the ability to select and specify the material. Through the experiential value of art, and through a certain virtuality of the image itself, it can be shown that what is ultimately generated is a hypothetical space in which the reality of a representation is, by necessity, a constructed and produced one, but in which events themselves can nevertheless be made known.[21]

Figures 5 and 6 (see Chapter 8, Figures 8.4a and 8.4b), wood engravings by James Davis Cooper after Briton Rivière, are the end result of collaborative work between Darwin and several artists including Joseph Wolf and Thomas William Wood. Darwin disagreed with the artists on certain issues such as the position of the head and wagging tail in the humble dog or the raised hair on the hostile dog, made notations accordingly on rejected drawings and in correspondence, and discussed them at length with Rivière. It is attentiveness to the *detail* of the animal's movements that allows the expression to become notable in the first place from Darwin's perspective. The expression as formulated then feeds into an analysis of the animal's emotional state. However, Darwin is not uniquely concerned with arresting a particular emotional state *per se*, but rather with the intersection of visual aesthetics and elastic physiological concepts. Darwin's intervention is an act of self-inscription into *Expression* that is contingent upon cultural observations and decisions.

Drawings are not the only instance in which Darwin inscribes himself into his imagery; he also had his photographic material altered and asked Rejlander—who provided, amongst others, images used to document fear, anger, pleasure and surprise—to modify the famous "Ginx's Baby" photograph (Figure 6.2). To produce "spontaneous" photographs of figures in motion in the 1870s presented considerable technical difficulties. The crying baby is thus regarded as one of the first photographs that circulated under the heading of "momentary images" and was independently marketed by Rejlander. The photographer seems to have been successful in capturing the child's expression at precisely the moment when his muscles and his face were in a state of the most extreme agitation. The image most often used by Darwin in *Expression*, however, is not a simple copy of the original. Rather, it is a photographic copy of a drawing made from a photograph.[22] Only minute *details* were changed in this process, but these changes were known to Darwin and regarded—for example, the infant's clothing and alterations to the hair and chair in which the child sits. The transfer from photographic image to drawing allowed Rejlander to alter, or rather, to *accentuate*, exactly those parts that Darwin attached the greatest importance to.

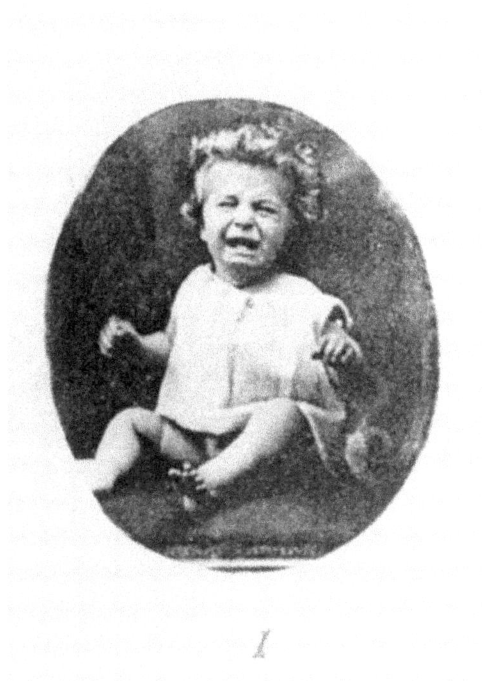

6.2 Plate I, Figure 1 of Charles Darwin, *The Expression of the Emotions in Man and Animals,* 1872. Mental Distress, or Ginx's Baby. Heliotype print. Oscar Rejlander and Adolph Dietrich Kindermann

With these interventions in mind, we can note that Darwin chose and altered photographs and drawings from life in order to create the impression of objectivity. The way in which Rejlander revised the photograph enabled him to re-emphasize those aspects of the child's expression that would otherwise have been lost in the process of reproduction. What is nevertheless striking is that he also introduced a number of *additions* to the composition, which effect changes in the child's environment—that is to say, alterations in the child's *habitat*. Following modifications, the child is found to be sitting in an armchair, with changes to the surrounding room as well. It is clear that these changes are intended to appeal to the viewer in terms of their own experience of everyday life by replicating an environment that Victorian viewers would feel intimately familiar with. Darwin's book thus is also attentive to settings that engage viewers in a unique manner, allowing them to relate to—and mentally to verify—the object of representation. I would, therefore, argue that Darwin was not only interested in developing a catalogue of expressive gestures that would become manifest *within* the image itself; more importantly, he also sought to *create* these expressions in the first place, namely *in* the viewers; that is, in terms of impressions based on the viewers' own perception and to elicit from the viewer the same emotions as those depicted in the images.

Darwin's intervention into the imaginary does indeed reveal itself as an act of inscription into *Expression* that is contingent upon cultural observations and decisions. It is also relevant to the concept of an iconic episteme concerned with the link between expression and emotion: namely, the internal convergence of imagination, memory, and image. In this sense, it is the imaginary itself that makes use of the image's suggestion of the internal in order to project itself upon what is viewed.[23] In light of Darwin's program, it is of some interest to consider current corporeal concepts such as recent discussion of the EBA—the extrastriate body area. This term denotes a neuronal system that is responsible for the visual perception of bodies. Discovery of the EBA by neurologists demonstrated that people do not only react empathically when they see another person engaged in an activity (according to the principle of mirror neurons, which explicitly relates to an organic body in motion), but that exactly the same neuronal responses occur when a subject is shown an *image* depicting a given expression or motion. The accentuations and modifications introduced into Darwin's images are intended to put the viewer into a state of emotional sympathy. Sympathy enables the viewer to empathize with what they see in the picture. Empathy, in turn, adapts itself to the bodily events that give expression to an emotion.

This intersection of nature and culture in representation touches upon factuality and fictionality, closely connected to the idea of *imaginatio* and *imago*. If it is possible to "read" an emotion, as a concrete sensation *from* the body by way of arresting it temporarily, it is equally possible to semanticize this emotion by way of devising a classificatory system of emotional states. It is in this way that emotions become factual, e.g. (re)cognizable, and decodable. Depictions of expressions, however, involve both dynamic representation and, in terms of the specific movement that is represented, function as a repository for cultural meaning. Expressions alone are fleeting, which stands in opposition to the "factuality" of emotions: through movement and sometimes cultural performativity, an expression partakes in the fictional. The factuality of an emotion does not necessarily stand in opposition to the "fictionality" of an expressive gesture; rather, both are contingent upon and interwoven with one another. In other words, in this case the experience of an expression often calls into play an encounter with the fictional (memory-based and cultural). In Darwin's representations of expression this "fiction" gestures towards a range of meaning contained in the term *fingere*, i.e. form, composition, design, acceptance, generation and imagination.[24]

Darwin, "the millionaire of strange and curious little facts"

The fictionality of cultural inscription can bring attention to the most minute movements or details—to what Darwin calls "accessories." In his introduction to *The Expression of the Emotions in Man and Animals*, Darwin duly emphasizes the importance of "accessories" as critical representational devices: "I had hoped to derive much aid from the great masters in painting and sculpture, who are such close observers. [...] The story of the composition is generally told with wonderful force and truth by skillfuly given accessories."[25] In his study, Darwin drew upon a wide range of sources of artistic representation. At first, he discounted their value for a truly comprehensive understanding of expressive gestures—to his mind, the strong affects often displayed in these sources were too heavily subject to aesthetic norm—but they nevertheless are, from the very beginning, constitutive of his enquiry into detail. By emphasizing the importance of "accessories" in visual communication, Darwin was considering questions of representation that made use of expressive gestures, paraphernalia in the environment, garniture, and other types of detail. Interested in the expressive aspects of the visual, he turned to the difficulty of his project at hand: "The study of expression is difficult, owing to the movement being often extremely slight, and of fleeting nature."[26]

If we accept Darwin's thought as conditioned by the representational methods of the arts, it follows that he has made a connection between movement and expressive details in that regard. This concentration on quasi "infrafigurative graphemes" points to the optical ciphers of movements of the body which are culturally recognized;[27] these transmit the sense, the meaning, of an expression that locates itself *in between* concrete figurations in order to support a believable transitory moment. This implied transitory moment is evident in, for example, Darwin's description of a dog that approaches another animal with hostile intent: "As he prepares to spring with a savage growl on his enemy, the canine teeth are uncovered, and the ears are pressed close backwards on the head."[28]

Insights and experiences from Darwin's study of the arts are reflected less in his concrete and explicit references to works of art than in his chosen methods of representation. By capturing a moment in the dog's sequence of movements, the affect is inscribed into the representation itself.[29] "Accessories," aides for Darwin in the representation of movement, are in themselves by virtue of their range of suggested meaning, mobile, even fluid.[30] We can thus state that Darwin is concerned with the study of—as Warburg puts it—*articulated accessories*, which allows both men to read the visual codes of emotion and energy without dissolving them into the seeming unity of the symbolic. If, as Georges Didi-Huberman has suggested, Warburg "s'etait engagé sur une voie inverse: révéler, dans l'unité apparente des symboles, la schize structurelle des symtômes" (was committed to a reverse path, to reveal, in the apparent unity of symbols, internal structural variables), we can extend this perspective to apply to both Warburg and Darwin: both are concerned with exposing the structural rupture of the symptoms themselves.[31] At the center of deliberations lies, once again, a dual goal: an attention to the intensity of internal *and* external movement, to be expressed as *motio* and *emotio*.

"Accessories" not only function as ciphers of movement in *Expression*, but also as an aide in decoding expressions, since they are both fundamental to Darwin's observations *and* carriers of passionately animated representation. At the same time, Darwin's observation of accessories as expressions of emotion enables him to formulate *how* a given habitus (habitual action) fares within the force field of habitat and heredity. Habits had preoccupied Darwin from the late 1830s on. In Notebook N in March of 1839 he wrote, "Emotions are the heredetary [sic] effects on the mind, accompanying certain bodily actions."[32] This highlights the close connection between neuronal and bodily linkages Darwin was considering and underscores his perspective on the necessity to conceive of and analyze a given habitus within the complex network of habitat—i.e. a subject's environment—and heredity—the subject's heritage.

In *Expression*, Darwin elaborated on the difficulties and complexities involved in decoding emotional expressions; such difficulties had roots in the interplay of "residual" and "associated" habits. Darwin brought up automatic or unwitting reaction as the result of habit, the fact that certain emotional states only arise from a specific habitus in the first place. Throughout his treatise, Darwin repeatedly invoked "actions," "movements," and "expressions;" terms which can denote both manifestations of the nervous system *and* individual gestures. Particularly where the emphasis is on nerve actions or activities, the neuronal system tends to assume the role of agent in its own right, thus, as it were, taking the place of the acting subject. In terms of habitualisation, actions are executed purely by force of habit that arise out of a complex system within which habit, adaptation to the surrounding system, and individualization converge[33] On this, Darwin speculates that "some physical change is produced in nerve cells or nerves which are habitually used" and that "this applies to the nerves of motion and sensation, as well as to those connected with the act of thinking" and "when our minds are much affected so are the movements of our bodies."

Conclusion

Warburg famously saw himself as a historian of emotional energies and their bodily expression. This notion of energetics underpinned his attempt to study the representation of movement in his historical materials. In comparing Warburg's and Darwin's interests in expression and representation, Foucault's notion of the ongoing energies of images can be invoked, which reads events and their representation not so much as static objects of knowledge, but rather as an ongoing process. A re-reading of Darwin from the perspective that I have considered might suggest that Darwin pursued the much broader and more comprehensive project of an image–act that engages the reader's cultural memory and emotional empathy, rather than exclusively the more narrow framework of the contextualization of an expressive movement.[34] This, in turn, enables one to consider Darwin's project within the constitutive interdependencies of art, literature, and culture.

Hence, a reading of the "seismographer" Warburg allows us to consider, in his interest in figures in motion, the interrelatedness of culture, art, and science as projected by Darwin and the epistemological traversing of established boundaries. As such, it also illuminates the intellectual space in which Warburg chose to position himself. Warburg's "Pathosformel" (which loosely translates as "form or formula of emotional style") thus is not only productive in its examination of the visual arts with their montage of specific cultural gestures and psychological features, or their historical ruptures of unexpected, exuberant figures (as in the bacchic movements of the nymphs of Botticelli with their dionysiac, antique heritage), but his "kunstgestaltende Gebärdensprache"[35] also furnishes us with a profound and wide-ranging arsenal of conceptual tools that help illuminate the significance of detail within images in the life sciences.

Notes

1. Quoted in E.H. Gombrich, *Aby Warburg: An Intellectual Biography* (1970), 2nd edn (Oxford: Phaidon Press, 1986), 72. Warburg's copy of Darwin's book is conserved at the Warburg Institute in London.
2. Aby Warburg: "Vom Arsenal zum Laboratorium." Typescript, with handwritten corrections, 16 sheets of notebook paper headed KBW (Kulturwissenschaftliche Bibliothek Warburg); all sheets dated: December 29, 1927. Warburg Institute, London.
3. Ernst Gombrich's biography on Warburg (see endnote 1), his many invaluable contributions notwithstanding, is an example of the kind of scholarship that portrays Warburg and his time as wholly influenced by evolutionary theory.
4. See Sabine Flach and Jan Söffner, *Emotionaler Habitus : Gefühle und Sinne zwischen subjektivität und Umweltrelation* (Munich: Fink Publishing, Wilhelm, 2011).
5. On the sub-phenomena of showing see Gottfried Boehm, *Wie Bilder sinn erzeugen: Die macht des zeichens* (Berlin: Berlin University Press, 2007).
6. Quoted in Hans Belting, ed., *Bilderfragen: Bildwissenschaften im Aufbruch* (Munich: Fink Publishing, 2007), 14. Translated passage by Alexa Alfer.
7. Reprinted in Michel Foucault, *Dits et Ecrits II 1970–1975* (Frankfurt am Main: Suhrkamp, 2002), 707–15 and Michel Foucault, *Dits et Ecrits IV 1980–1988* (Frankfurt am Main: Suhrkamp, 2005), 294–302.
8. In this case, the models, one of which is the photographer himself, have posed to accommodate Darwin. Darwin acknowledges this as follows: "The figures 1 and 2…are fairly good representations of men simulating indignation," and of the two figures feigning helplessness Darwin writes, "Mr Rejlander has successfully acted the gesture of shrugging the shoulders." (Charles Darwin, *The Expression of the Emotions in Man and Animals*, ed. Paul Ekman, 3rd edn (Oxford: Oxford University Press, 1996), 242, 265). Darwin himself, when viewing these figures together, initiates a possible scenario, writing later in his text, "…we may imagine one of the figures on the left side having just said, 'What do you mean by insulting me?' and one of the figures on the right side answering, 'I really could not help it.'" (272). I thank Barbara Larson for pointing out this passage.
9. See Ulrich Raulff, *Wilde Energien: Vier Versuche zu Aby Warburg* (Göttingen: Wallstein, 2003), 348.
10. See Boehm in Belting, *Bilderfragen*, 82.

11. Barbara Stafford, "Neuronale Ästhetik: Auf dem Weg zu einer kognitiven Bildeschichte," in *Iconic Turn*, eds Christa Maar and Hubert Burda (Cologne: Dumont Verlag, 2004), 107.

12. See David Freedberg, "Empathy, Motion, and Emotion," in *Wie sich Gefühle Ausdruck verschaffen: Emotionen in Nahsicht*, eds Klaus Herding and Antje Krause-Wahl (Taunusstein: Driesen, 2008).

13. As shown by science historian Michael Neve. On this, see Julia Voss, *Darwins Bilder: Ansichter der Evolutionstheorie*, (Frankfurt am Main: Fischer Taschenbuch Verlag, 2007).

14. Charles Bell quoted in Philip Prodger, "Illustration as Strategy in Charles Darwin's *The Expression of the Emotions in Man and Animals*," in *Inscribing Science: Materiality and Text in Scientific Communication*, ed. Timothy Lenoir (Stanford, CA: Stanford University Press, 1988), 140.

15. Quoted from Voss, *Darwins Bilder*, 317.

16. In terms of heredity and habit, consider the baring of teeth, originally threatening and no longer useful, as an expression that still occurs. This was the subject of one of the photos in *Expression*, as figure 22, described by Darwin as "a lady, who sometimes unintentionally displays the canine on one side."

17. See Klaus Kruger, "Die Bild als Palimpsest," in *Bilderfragen*, Belting, 140.

18. Thomas de Quincey, "The Palimpsest of the Human Brain," in *The Collected Writings of Thomas de Quincey*, ed. David Massoned, 14 vols (Edinburgh: Adam and Charles Black, 1889–1890) 13, 340–49. On this also see Freud's "Wunderblock," in which he conceives of the human perceptual apparatus as a dynamic structure of the formation, storage and sedimentation of human experience. Originally published in *Internationale Zeitschrift für Psychoanalyse*, 1924, reprinted in *Gesammeltewerke*, 13 vols, 1, 1–5.

19. Raulff, *Wilde Energien*, 43.

20. See Giorgio Agamben, "Wiederholung und Stillstellung. Zur Kompositionstechnik der Filme Guy Debords," in *documenta documents* 2, ed. Documenta GmbH (Ostfildern: Cantz, 1996), 68–75, esp. 73.

21. See Sabine Flach, "'It's not easy being green!': Schnittpunkte von Kunst, Medientechnik und Naturwissenschaften am Beispiel der Transgenic Art," in *Konstruierte Sichtbarkeiten: Wissenschafts- und Technikbilder seit der frühen Neuzeit*, ed. Martina Hessler (Munich: Verlagsort, 2006), 281–303.

22. On this history see Phillip Prodger, *Darwin's Camera: Art and Photography in the Theory of Evolution* (New York: Oxford University Press, 2009), 103–30.

23. See Boehm in Belting, *Bilderfragen*, 80.

24. See Sigrid Weigel, *Genea-Logik: Generation, Tradition und Evolution zwischen Kultur- und Naturwissenschaften* (Munich: Fink Publishing, 2006).

25. Darwin, *The Expression*, 14.

26. Ibid., 19.

27. Raulff, *Wilde Energien*, 145.

28. Darwin, *The Expression*, 56.

29. Conference paper by Elisabeth Strowick, "Ausdrucksbewegungen. Zum Verhältnis von Poesie und Naturkunde bei Durs Grünbein, Ian McEwan und Charles Darwin," Technische Universistät, Berlin, Institut für Literaturwissenschaft, November 2007.

30. Raulff, *Wilde Energien*, 28.

31. Georges Didi-Huberman, *L'image survivante: Histoire de l'art et temps des fantôme selon Aby Warburg* (Paris: Minuit, 2002), 493.

32. Darwin, *Charles Darwin's Notebooks, 1836–44: Geology, Transmutation of Species, Metaphysical Enquiries*, transcribers and eds P.H. Barrett. P.J. Gautrey, S. Herbert, D. Kohn and S. Smith (Ithaca, NY: Cornell University Press, 1987), 581.

33. Weigel, *Genea-Logik*.

34. Ibid. and Didi-Huberman, *L'image survivante*.

35. Aby Warburg, *Der Bilderatlas MNEMOSYNE*, ed. Martin Warnke (Berlin: Akademie Verlag, 2003), 2nd edn.

7

On mimicry in Darwin and Surrealism

Jan Söffner

Prologue

After a long day's visit to the bird park near Walsrode, Germany, my partner and I wandered by some of the less spectacular cages. In one of them, a large dove soared above a recreated landscape. It seemed to be at once annoyed and bored by a blackbird-like species, whose fluttering presence in the cage appeared to me at first little more than a complement to the majestic dove. When we approached the cage, we were alone, and so it was a bit unsettling to hear human voices. They said in a completely disinterested manner: "Guck mal die Taube" ("Look at that dove") and "Ja" ("Yup"). For a moment I doubted my sanity before I finally understood that the small "blackbirds" were mynahs—and that it was they who "spoke."

The vocal mimicry of mynahs has been well studied. The adaptive value of mimicking seems to lie in a hierarchical function for flocks (vocal mimicry there is a sign of dominance).[1] It is an open question exactly how this kind of imitation by mynahs relates to reward.[2] However, despite the fact that mynahs tend to mimic human voices predominantly in captivity only, there is still some evidence that they can make use of other animal's alarm cries and chase them away (and steal food), even though they most likely do not act "on purpose" (which would include conscious planning), but rather respond spontaneously. This seems to relate especially to primates. Wild mynahs have been spotted mimicking the alarm call of gibbons and langurs.[3] In the case of the human primates at Walsrode, it always appeared to "work." As visitors approached the cage and the birds uttered a specific cry ("Look at that dove"—"Yup"), the humans always left.

This scenario might resemble one of the many semiotic anecdotes in which a kind of communication apparently works just as well for either "speaker," but, in reality, "communication" is based on profound misunderstanding due to different systems of signification and a different use of arbitrary signs. And, indeed, when considering mimicry and its impact on cultural phenomena, a semiotic approach might be important, even when dealing with cross-species interaction. Nevertheless, I think that in this particular case, a semiotic scenario like this would be too facile. Indeed, most semiotic models of mimicry are too limited to address the complexities of what mimicry is. Moreover, cross-species interaction and any aesthetics evolving from it cannot be considered without the impact of Darwinian theory for both cultural history and the ways of describing natural phenomena. This intricate relation of semiotics and Darwinian theory in describing the aesthetics of mimicry is the topic of this chapter. I wish to state at the outset that one should be cautious in describing mimicry in semiotic terms only since this overlooks a crucial dimension of embodied interaction.

Let me trace what would be the most convincing of the traditional models—and here I am drawing upon Peircian semiotics. This would posit that the simple alarm cry of a gibbon would be an index. There would be a non-arbitrary link between a qualitative experience of perceived danger and the cry. The cry would entail a *quale*, a kind of a sensori-emotional quality, a sound experienced as fear. Mimicry would introduce a second dimension into this simple figuration. It would enter the domain of representation by likeness: it would replicate the gibbon cry and no longer be in itself an index of fear and danger, but just a simile of this index. By mere likeness (instead of an indexical relation), a mynah's cry would be a mimetic *representation* of an index, not an index itself. And the fact that this second state applies to only one of the species would be precisely the reason why mimicry works. For a gibbon, who knows alarm cries only as indexes, the cry would therefore *be* an alarm call—even though for a mynah, it only *mimics* an alarm call. In Peircian terms it would be an "icon" instead of an "index." A word spoken by a human being, according to this model, would instead offer another scenario. The word would not mimic anything specific, but it would just arbitrarily refer to something else in its absence. Even the mimetic bond would be lost: communication would be a fact of "symbols." To apply this model to the situation described, the (mis)communication would be understandable like this: A bird mimics a human word—and thereby it lets this very word re-enter the state of mere likeness, of iconicity, instead of symbolic signification. While a human being understands a word for what it *means* (as a symbol), a bird imitates it regardless of what it refers to in a human language (as an icon). Now, evidently, this *is* a model for describing how mimicry works. So why should it be too facile? The reason why is that this model is unable to account for the phenomenology of mimicry on the evolutionary and non-symbolic level where it is really mimicry.

One might ask whether a mynah really does have icons. If it *does* simply by replicating a sound, this fact does not take phenomenology into consideration since, if we accept Thomas Nagel's approach and ask what it is like to be a mynah, it is unlikely that these birds mimic merely to re-present a cry.[4] If there is any intentionality involved—that is, if mimicking is *about* anything, it is about chasing away, not about producing likeness for representation. If, instead, the mynah really had a "plan," the imitated sounds would even have a symbolic meaning as well. They would express something like "gibbon-like animals go away." I would argue that the mynahs do not really seem to *intend* to produce an "iconic" likeness because, as stated above, this would also imply conscious planning and the production of building mental scenarios or models of the future. It also would imply the possibility to "lie" and to engage in a kind of "theory of mind" about other animals. A bird mimicking an animal cry probably is not lying, as a human hunter would be if mimicking an animal cry. Instead, the mynah most likely cries *with* the cries present in a given situation, i.e. it does not try to convey a "message," but it chases away in the process of an embedded response. The interaction present in vocal mimicry is not "meaning"-based, but enaction-based. For the mynah, mimicking an alarm cry does not seem to be an attempt at "speaking a different language," but of acting within a given situation. Human mimicking, i.e. consciously "lying" (conveying false messages) in order to deceive somebody else's (the animal's) "mind" is different from that. As Jerome Lewis has pointed out with relation to the Mbendjele language, which includes the mimicry of animal cries in a way unlike any other known human language, Mbendjele hunters know very well that animals "have extreme difficulty recognizing that the hunter is lying."[5] It is probable that only a human being is able to use the cry as a symbol (part of a language). That means that in human imitation of animals, mimicry becomes something different from the enactive animal mimicry; it becomes representational mimesis, an imitation of something that represents in the context of an absence and that can work as an arbitrary symbol in a larger system of language. If it becomes "iconic" in this sense, mimicry ceases to be mimicry.

Accordingly, I think the weakness of the semiotic model is that it draws upon *qualia*—i.e. discrete states felt *by* a kind of equally discrete consciousness finding itself vis-à-vis its environment. If we follow this model, we implicitly accept the presence of *observers* (animals / humans *for which* the cry appears to be this way or that) and *objects* of observation (sounds appearing as discrete "cries of alarm"). The semiotic model thereby *does* account for a model of perception as a construction of internal states *referring to* something *external*. However, it fails when accounting for perception as a part of action and for meaning (or meaningfulness) as emerging from embodied activity. Or, to put it differently: it makes meaning a fact of *expression*, not a fact that is intrinsic in action and emerging from it. It accounts for representation instead of participation.

Mimicry, though, in an evolutionary sense of the word, is not about (passive) perception of the replicated acoustic or visual stimuli. It is about perception as action or "action in perception," to use Alva Noë's term.[6] This means things are not about perceptive *qualia* (e.g. a bird simply seeing the *quale* of an "eye" on the wings of a peacock moth); they are about "affordances"—to use James J. Gibson's term—i.e. they are action possibilities or rather complements of actions latently existing in the environment.[7] They entail enactions guided by emotions, of which the perception is just one aspect (the bird avoids the peacock moth just as a human driver would avoid a pothole by steering away from it, rather than by first constructing a conscious *quale*). The problem of a semiotic approach to mimicry is hence that it would make a fact of signified meaning out of what is an enaction, and that it would conceive of meaning in terms of *qualia,* i.e. some kind of subjective ("inner") states of mind to be "expressed," something conscious opposed to an outer world, instead of being a part of acting and being in this world.

I wish to argue that there is a strong "Cartesian" anthropomorphism in this view, and that this anthropomorphism is due to representationalism. If something present can "express" or "stand for", i.e. represent something else, then the represented object is present only in an immaterial way: as a state of mind working upon a state of body, but not emerging from the state of body or even being a part of it. Semiotic theory accordingly entails a categorical distinction between *qualia* (as a somewhat immaterial "state of mind") or at least *meanings* and embodied activity.[8] The mere fact of establishing something "given" as an object and distinguishing it from what it means leads to an inability to describe meaning as intrinsic in action. Even the Peircian model thereby does not account for "perception in action," and by failing to do so, it also fails to account for the Darwinian dimensions of mimicry and also for its phenomenology. This is my main thesis. In what follows I wish to: (a) elaborate on why I think the (traditional) semiotic approach to mimicry to be misleading; (b) trace the consequences of terminating this semiotic premise; and (c) discuss what implications a different, more Darwinian model of mimicry can have for aspects of human art, especially Surrealism.

Ceci n'est pas un *x*

One of the most famous paintings by a Surrealist seems to be oddly un-Surrealist and rather banal at first glance. I am referring to René Magritte's *La trahison des images* (*The Treachery of Images*), a depiction of a pipe with the handwritten subtitle: *Ceci n'est pas une pipe* (*This is not a pipe*) (Figure 7.1). The issue seems to be about mimetic art representing an object in the world. Magritte himself stated that quite naturally the depicted pipe is not a pipe—and to make this issue clearer, he pointed out the impossibility of putting tobacco into it. This account seems to make nothing but an optical riddle out of the image—and this riddle has a simple solution. It goes: well, true it is not a pipe—it is just a *representation* of a pipe. Or, to put the same solution into

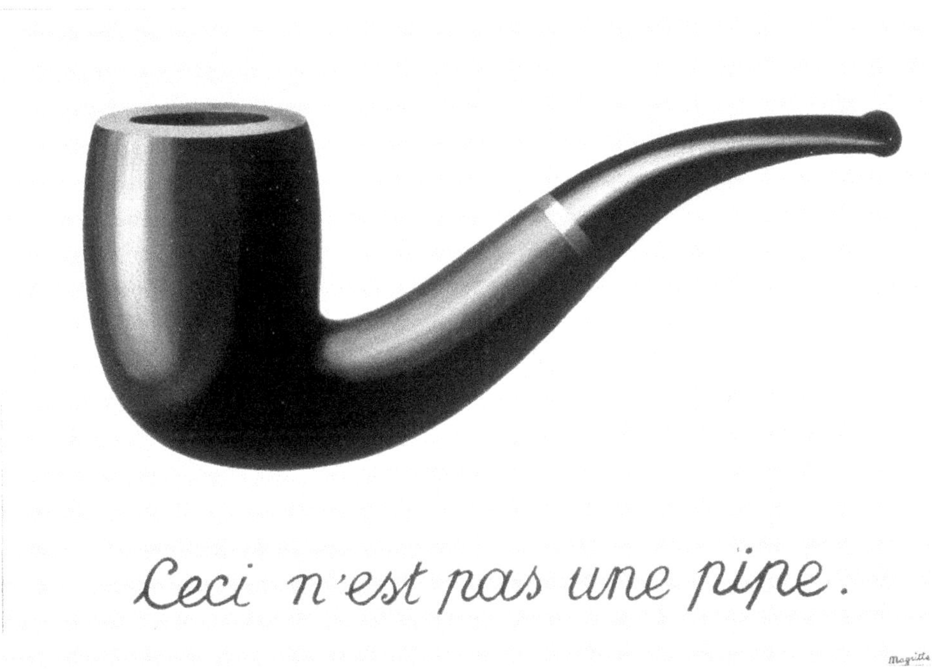

7.1 René Magritte, *La trahison des images (Ceci n'est pas une pipe)*, 1929. Los Angeles, CA, Los Angeles County Museum of Art (LACMA). Oil on canvas, 60 × 81 cm, © 2011. BI, ADAGP, Paris / Scala, Florence

Peircian terms: it is an "icon" whose likeness makes it refer to an absence. The pipe of the image, by means of mimesis, seems to *represent* a pipe, which itself is not present.

Something similar could work for my experience at the bird park, but here the issue would be about symbol and icon. I had to solve the riddle that "Look at that dove" was *not* a sentence, but merely an imitated sound, or icon. Once understood as a riddle, the issue would be analogous to a sentence uttered about an inscribed letter x: "This is not an x," making the reader try to grasp what an "x" is, if it is not a letter, but perhaps a cross of St Andrew (an icon) or even some mere impression of a black shape on a white background. The problem of this answer to the riddle is that of any such solution: the answer makes the observed phenomenon a riddle. A riddle, though, cannot account for the fascination that survives its solution. Therefore, the question remains, can this simple consideration concerning representationalism and mimesis really be that which gives Magritte's image such lingering interest. What makes people enjoy looking at this image over and over again? Are viewers so obtuse that they constantly want to be reminded that they once solved a simple riddle? Or is there something more to their observation?

Michel Foucault has discussed the threshold between a calligram and a referential text, between the representing function and the mere appearance of the image. He was intrigued by the destabilizing effects produced by the combination of both—in other words, the destabilization of two traditional principles. On the one hand, there is the principle of the dissociation of linguistic signs (being without resemblance, i.e. "symbols" according to Peirce) and plastic elements ("icons"). And on the other hand, there is the pictorial equivalence of resemblance and affirmation—which, since affirmation is a linguistic issue, reintroduces the linguistic order into the pictorial order. The painting would, therefore, be a work of epistemological figurations, destabilizing a given semiotic order of how to grasp images. In his eyes, the focus is on destabilizing a given way of organizing the relation between symbolic and iconic representation.[9]

A destabilizing momentum, an impulse obscuring a common way of perceiving iconic and symbolic representation, could also be in line with André Breton, though, who opened his treatise on *Le surréalisme et la peinture* (1928) with the phrase: "L'œil existe à l'état sauvage"[10] and who subsequently argued that perception does have a state of seeing what one is told to see or said to see—but that this state of "realist" perception is not all there is to it. There also exists, he argues, a "language" of images that has nothing to do with these mediated *meanings*—but rather consists in a raw and immediate kind of visual perception.

Breton, moreover, uses the word "spirituality" (*spiritualité*), thereby giving rise to doubt about the sensuality. In addition, Surrealism tends to focus on the unconscious and automatic rather than on consciousness. I think that we can gain an idea about what he might mean by having a look at a passage of *Nadja,* exposing a similar theory of the image with regards to the work of Giorgio de Chirico. Breton writes: "Chirico a reconnu [...] qu'il ne pouvait peindre que *surpris* [...] et que toute l'énigme de la révélation tenait pour lui dans ce mot: surpris."[11] (De Chirico recognized that he could not paint anything but *surprise* [...], and that for him, all of the enigma of revelation was contained in this word: surprise.) And he quotes the painter himself, stating that for him, his art was about a strange kind of likeness or resemblance between phenomena that could not be entirely distinguished from one another, such as: "l'image en rêve d'une personne déterminée et cette personne réelle. C'est, en meme temps ce n'est pas la même personne" (the image of some person in a dream is like the real person. It is, and at the same time it is not, the same person).[12]

Likeness in this Surrealist account (which takes de Chirico as a starting point) is not about iconicity. The first quote (on de Chirico) is not about what pictures represent, nor what one feels about them, but rather what one experiences as the uncertainty in the act of beholding the image, a surprise not *about* reality, but a surprise "à l'etat sauvage." The "surprise" replaces what otherwise would have been either reality or the enigma about it.

The second quote (by de Chirico), instead, indicates that imitation in Surrealist images requires a state where likeness is not apt for *referring* to something or somebody by *resembling* this something or somebody, but rather being and, at the same time, not being, this somebody. This is obviously not possible in terms of producing objects that are like other objects. But it is possible in participatory *enactions* and *interactions* that require or consist in processes of assimilation—e.g. if one considers the vocal mimicry of a mynah not as a reproduction of stimuli, but rather as an act of *taking part* in an ongoing interaction with humans. Perhaps one can understand it in a similar way, if Surrealist images produce a kind of likeness that does not distinguish if something is *like* something else, if it *is* that other thing or if it neither refers to nor is that thing, but is something beyond this very question. If it were that way, then likeness does not bridge the gap between one item and another, but rather unsettles the boundary between one object and another, as well as the boundary between subject and object. Indeed, Breton searches for likeness beyond the distinction of separate items, and accordingly, beyond the very question of what these items could refer to. The likeness that he would call Surrealist is neither referential, nor self-referential, nor a-referential. It does not make one object refer to another, but it rather makes them appear to be the same object and not the same object at the very same moment.

This may sound rather esoteric, but let us now turn back to Magritte and the attempt to grasp why theories of artistic mimesis (understood as representations based on likeness) fail to give an account on the aesthetics of this image. I would argue that in the case of the Magritte painting, it is not about unsettling *reality* (the somewhat intellectual process of comparing a mental model of the world with the sensory input of information, the mimetic representation); it is rather about unsettling *perception* and its habituation (and here I would agree with Foucault, although I believe he remained too semiotic and representational in his approach). Therefore, it is closer to (natural) mimicry than (artistic) mimesis. If Magritte, indeed, *had* depicted nothing more than a riddle about representationalism, and if the fascinating point of this picture could be accounted for by solving this riddle, then it would be explainable in semiotic terms alone. But it cannot be. And the unsettling point, in my opinion, is that the picture plays with the possibility and impossibility of something that is *always* relevant in semiotics: negation.

In perceiving a representation of a pipe *as a representation*, there is the option of "as if" and of negation. One can say that it is "as if" one were "really" seeing a pipe—but at the same time, one can know that it is not. This is what makes it an artistic mimetic representation. Of course, at first glance, natural mimicry can seem to be representational as well. Once one has spotted the "disguise," one can say that watching a hoverfly is *as if* watching a wasp. One establishes an iconic relation of likeness between these two species and enjoys the deception by "suspending disbelief," to use the words of Samuel Taylor

Coleridge. Nevertheless, as explained above, this is exactly how mimicry (in nature) does *not* work. If a bird had to form a concept of and belief that what it was seeing was a dangerous wasp with its affordance of being avoided, it probably would never be fast enough to get to the point of avoiding the hoverfly. And if a beholder had to suspend disbelief in viewing the Magritte, they would not be unsettled by the phrase "Ceci n'est pas une pipe." In order to unsettle human perception, both mimicry and Magritte's painting need a stratum of perception without this option for "as if" and for negation—one that interferes with the "semiotic" perception of seeing the pipe as a mere representation.

Troubles without "as if"'

One question for understanding the dimension without "as if" is: does the bird really discern a hoverfly *as* a wasp, or even does it really discern a wasp as a wasp—even though it has no word for wasps, nor (probably) conscious concepts about them? According to Nagel's approach, one can give a tentative answer to this question. We know that the enactive relation of birds towards their prey depends on the coupling of senses, emotions, and actions. Emotions, though, are evolutionarily successful because they are intrinsic to action—they do not have to build mental models of the surrounding world, but work with affordances in a different way. This is true for humans too. When seeing a snake, one *first* jumps away, *then* feels the fear and *only then* one makes up one's mind that this fear was about a snake (indeed there appear to be two different circuits in the brain processing this perception).[13] If one had to make up one's mind first, there would be a lot more snakebites in the world. The interesting issue about this automatism is *first* that it does not even imply an indexical relation (the snake does not *mean* the danger of being bitten, but jumping away is a part of seeing it—we are addressing affordances as perception in action, not *qualia*); and *second* only later, when finally forming the mental model does the question of deception arise—and with it arises the option for negation. For the perception in action itself it does not matter if we have seen a (venomous) coral snake, a (harmless) scarlet king snake mimicking it, or a rubber dummy mimicking one of the two. It is probable that a similar phenomenology also applies to birds seeing wasps or hoverflies.

The fact that the state without "as if" applies to humans too does not imply the switch from one state into the other, however. It is not the case that the kind of perception involved in jumping away *ceases* in the moment of cognition. Rather it remains present. The question is not about thresholds, it is about different states or stratums flooding the organism contemporaneously in enacted perception. And the same is true for the phenomenology involved. The question is not whether this state or that state is "conscious." It is about latency and manifestation. It also means that we have to take the *co-presence* of (at least) two relations to the world into account—a semiotic relation of indexes, icons and symbols *and* a non-representational relation beyond any option for "as if" or for negation—a relation in which we can see a pipe as a representation and one in which we cannot.

In the light of negation not being an ultimate possibility, one can apply Freud's statement that there is no option for negation in what dreams show. If, using his theory, a dream "wants" to say that things are *not* about a pipe, it (a) thereby *is* already staging a pipe, and (b) cannot negate this pipe except by marking it as somehow "unreal"—e.g. by accompanying it with a feeling of strangeness—as do the "surprise" constellations in a de Chirico (in Breton's theory) and as does Magritte's sentence, "Ceci n'est pas une pipe" in the *Trahison des images*. This, in my view, seems to be the very fact which Breton is talking about when he considers the importance of surprise rather than representation; and that he considers likeness beyond the option of being likeness with something or somebody. Without negation, likeness is something very basic and simple. It lies in the impossibility of distinguishing something being there *as* what it is from something being there *like* something else. It is about the impossibility of distinguishing likeness and as-ness.

The treachery of images and the treachery of natural mimicry accordingly share this very constellation of simultaneously being apt and unapt for negation. They have a dimension where there is no option for negation (where likeness is not really about hoverflies or wasps, pipes or representations, but about emotions and enactions). But on the other hand, this very dimension seemingly interferes with and contradicts the deceptive object that can be solved like a riddle of negation (where the hoverfly is *not* a wasp and the representation of a pipe is *not* a pipe).

My point has to do with the co-presence of both states and the fact that the non-representationalist account remains latently present even after having solved the riddle or spotted the "disguise." Observing the hoverfly or the painting many times over seems to activate some kind of "perception in action," which is independent from what a traditional semiotic account would describe—a stratum of perception that has nothing to do with enigmas or riddles—and therefore cannot be "solved." The treachery is, accordingly, a fact of not being able to avoid deception, even if one has already intellectually dismantled it. It is beyond the distinction of "as" and "like." The phrase "Ceci n'est pas un pipe" is, therefore, itself a kind of treachery hiding the fact that, at the same time, the pipe *is* a pipe.

Let me draw some first conclusions about what is missing in the semiotic model of mimicry as discussed above. I think that there are three issues that matter here. The first two have to do with its lack of phenomenological depth, whereas the third has to do with evolution:

1. The semiotic model of mimicry does not account for the co-presence and interferences of representational and non-representational states of perception. It rather describes—or better, states—the theoretical differences between three states of signification on an ontological level. It describes what they *are*, not what they feel like or how they work.
2. Since the semiotic model presupposes the three states of index, icon and symbol to be distinct from one another, it cannot offer a convincing model for the phenomenology of passing from one state to the other—and, thereby, going back and forth. This going back and forth is, though, exactly what "iconic" illusions (e.g. the treachery of images at the same time distinguishing and not distinguishing likeness and as-ness) are about (and indeed, something similar applies to the "symbolic" illusion of the cries of the mynahs being and at the same time not being words).
3. Presupposing an observer for whom objects are objects means to presuppose some kind of basic representationalism or at least proto-representationalism (where presence would be the presence of an object *as an object* and *for* somebody *as a subject*). Either an immaterial signified or an immaterial mind is opposed to embodied enaction and, therefore, no model of perception in action is possible in semiotics, which is crucial in describing the evolutionary aspects of how mimicry works. Drawing on Breton's and de Chirico's concept of likeness, I suggest that we need a different concept of imitation for understanding mimicry.

Darwin and the *ergon* of mimicry

Darwin does not follow a semiotic paradigm. His theory is about adaptation instead of signification. Perception is *only* considered in so far as it is a perception in action. In *On the Origin of Species*, Darwin accordingly discusses mimicry as a fact of analogy shaped by adaptation and natural selection, just as elsewhere he would describe a fish's tail as a fact of hydrodynamics shaped by adaptation and natural selection. There is no categorical difference for him whether adaptation refers to a merely physical habitat of water or whether it refers to a biological one with other animals present. Of course, there is a continuum between the two. Regardless, it is only to our own semiotic eyes that the phenomenon of mimicry seems to be like a dissimulation or disguise, e.g. when it seems like the hoverfly *hidden behind the appearance of* a wasp. *We* are trained in distinguishing seeming from being, aesthetic appearance from reality, an "as if" from an actual state. Evolutionary mimicry is not an imitation of some object, animal, or property; it is the adaptation to an embodied and enactive *relation*. Like the adaptational functionality, e.g. of a whale tail, the similarity of a hoverfly and a wasp is *non-representational*. And this is why Darwin could discuss mimetic resemblance (for example, in Batesian mimicry, wherein the species preyed upon—as in the case of the hoverfly—exists as it does because predators have eaten those that least resemble wasps) as an example of the larger principle of adaptation. The implicit point is *functional* similarity (the look that repels predators), as opposed to *referential* similarity (the look that, by means of similarity, refers to wasps). If it were referential similarity, it could not work: predators, like humans, would understand the difference between an as-if wasp and a real one.

Darwin himself, nevertheless, *does* call mimicry a form of imitation, so how can imitation be conceived of without referring to any concept of representation as well? In order to make this point, I wish to draw on some aspects of Plato's theory of mimesis, giving some reasons why I consider it to be non-representational as well.[14] In his *Republic,* Plato describes the difference by referring to a comparison of a bed made by a carpenter and a bed made by a painter. Both of them *are* beds, in Plato's eyes. He writes without a concept of representation. There is no such thing as a painting just representing or referring to a bed. And accordingly, his concept of mimesis neither allows for "as if" or for negation.

The argument, indeed, is not about representations or "effigies" (as has often been argued). It is about whether the *ergon* ("work" or "functional meaning") or *ousia* (being) of a bed is realized, or whether it is just simulated. That is, whether you can sleep in it—or whether you cannot. And indeed, this is why Plato found it plausible to argue that the *ergon* of the bed is what beds are about. The *ergon* of beds is some kind of sleep-in-hood—and this is what the bed of a painter lacks. Plato might have said in regard to Magritte's *Trahison des images*: This *is* a pipe. But it is a bad one. You cannot even put tobacco into it, and hence it has no *ergon* or *ousia*. Considering adaptation as a process of improving functional relations with the environment, Darwin's concept of mimicry somehow shares Plato's idea of non-representational mimesis as a kind of produced likeness with the realisation of *erga*. As Plato's carpenter forms the sleep-in-hood of beds in a more or less perfect way, Darwinian evolution forms the swim-with-hood of whale tails in a more or less perfect way—and it also forms the look-dangerously-hood of hoverflies in a more or less perfect way.

But what about the painter's mimesis and what about natural mimicry? Here the differences between Plato and Darwin are clear. For Plato, the fact that you cannot sleep in a painted bed or smoke a painted pipe is enough to say that a painted table has no *ergon* at all. It is only realized for the visual senses and the emotions, not for purposeful interaction as a whole; and hence the table or pipe of a painter is just a simulacrum (*eidolon*). It is a fact of bad mimesis. In Darwin, though, the relation with the senses and emotions do have their *erga*. Evolution does not know bad mimesis. Or, more concretely, the simulacrum of a wasp, Darwin argued, has its own adaptive function.[15] It has, to use the Platonic term, the *ergon* of keeping predators away.

And, of course, it is the fault of Plato's ontology that he failed to understand that aesthetics as such can have their specific *erga* as well (which is definitely the case in *mimicry*).

The "two cultures" of mimicry

Most influential biological concepts of mimicry—at least those that have proven to be sustainable until now—evolved in contact with Darwin, both on a personal level and on a theoretical one.[16] Henry Walter Bates, who was instrumental in the development of the theory of mimicry protecting defenseless species by adopting some traits of dangerous species, was in contact with Darwin. So was Alfred Russel Wallace, who is known for his discovery of the function of "warning" colors keeping predators away from venomous animals and those mimicking them. And so was Fritz Müller, who first described the fact that different unpalatable species share the same warning colors (and thus mimic each other), so that each new generation of predators does not have to learn new signals all over again, and the death toll of those animals killed "to find out" would be smaller. And an equally "Darwinian" thinker was Edward Bagnall Poulton, who developed the first large-scale classification of different kinds of mimicry. In each kind of mimicry, the *ergon* is not really what "deception" would mean if understood as misguiding the predator's *cognition* by making it "believe" this or that, but it is rather a misguidance of *enactive* features of perception (keeping the animal away). There is no simulation of a different "object," nor a disguise of the animal's or plant's own "identity" in mimicry at a natural stage. These concepts would require a kind of mental representation or subjectivity of the "other" animal—and these categories only come into play if we accept a semiotic approach instead of one guided by adaptation. The important issue is that as in Darwinian functionality, the *ergon* of mimicry is not one due to observation, but to relations and sharedness. There is, of course, an apparent and vividly discussed difference between mimicry taking place as "blending in" a given and unspecified environment on the one hand—and a "winnowing out," where mimicry leads to similarities with a concrete and specified other species. Nevertheless, in the shared interaction with predators, i.e. in the act of keeping them away by interacting with their sensuality, both rely on shared interactiveness. Therefore, even in these latter cases, imitation or likeness does not lead to becoming an "icon" *for* the predator (in the Peircian sense). The "simile" does not *refer* to anything, nor does it represent anything—it is not even "representation-rich";[17] and, therefore, it is not even a masquerade in the full meaning of the word.

A non-semiotic definition of mimicry, though, could not please those thinkers who, in one way or another, had engaged in romantic meditations on a meaningful nature and who wished these ideas to function within the same paradigm as culturally construed meaning. A natural phenomenon which, to human eyes, seems so much like a human phenomenon of playful disguise and dissimulation was not very likely to be interpreted in a Darwinian way by those who had been working with these semiotic strategies for such a long time. Those who wanted to see semiotic principles and a *qualia*-based perception, indeed, tried to make "meaning" the core of natural mimicry. Vladimir Nabokov, for example, in his novels *The Gift* and *Pale Fire* made his characters engage in a serious argument against natural selection in mimicry. He seemed to favor a view of the world in which all these ruses and disguises were only observed as such by humans (which is probably true) and that, therefore, the world seemed to be "intended" for this very human gaze (which is instead very improbable).[18]

What is more important for this essay, though, is the case of the Surrealist Roger Caillois, who, in his famous *Mimétisme et psychasthénie légendaire,* favored a magic law of Likeness, namely the "*attractio similium* [...] *le semblable produit le semblable*" (the similar produces the similar)[19] over natural selection and adaptation. This very magic, in his eyes, was equally present in natural mimetic phenomena and those of certain mental states. Perhaps with reference to Gabriel Tarde,[20] Caillois conceived of imitation as a fundamental law of both nature and culture, needing no function or *ergon;* nor did that law need any further explanation, except a phenomenality of effecting the dissolution of the subject or ego, its disappearance in its habitat. Moreover, the likeness produced by *willful* imitation[21] is, in his eyes, almost by definition already a representation—be this in the form of a "véritable photographie, mais de la forme ou du relief" (true photography, but of form and profile) or in the form of a mental phenomenon equally linking concepts of likeness and reference, so that thinking of mimicry in a semiotic sense remains possible despite the disappearance of the subject, despite the loss of its self in its habitat. The ego disappears as an index and as an icon, but it does not as embodied existence. And without this semiotic premise, it would be impossible to explain how becoming invisible could be equated with a cessation of the ego. Outside of his theory, soldiers, for instance, cannot really be said to lose their ego when wearing camouflage.

Indeed, there was another type of mimicry in culture as well, which was not so much focused on semiotics (and its—post-romantic—phenomenology), but rather on the *shared ergon* of camouflage. The best example of this culture was perhaps the painter and hobby-biologist Abbott Handerson Thayer, who discovered the mimetic function of the sometimes dazzling effect of disruptive colors (as in the stripes of a zebra or a tiger), which "fragmentize" the view of an animal on both a light *and* a dark surface—instead of making it invisible in front of one background and highly visible in front of the other. Thayer tried to propose "ruptive coloration" for camouflage strategies during the First World War; after a series of disappointments his proposal was accepted.

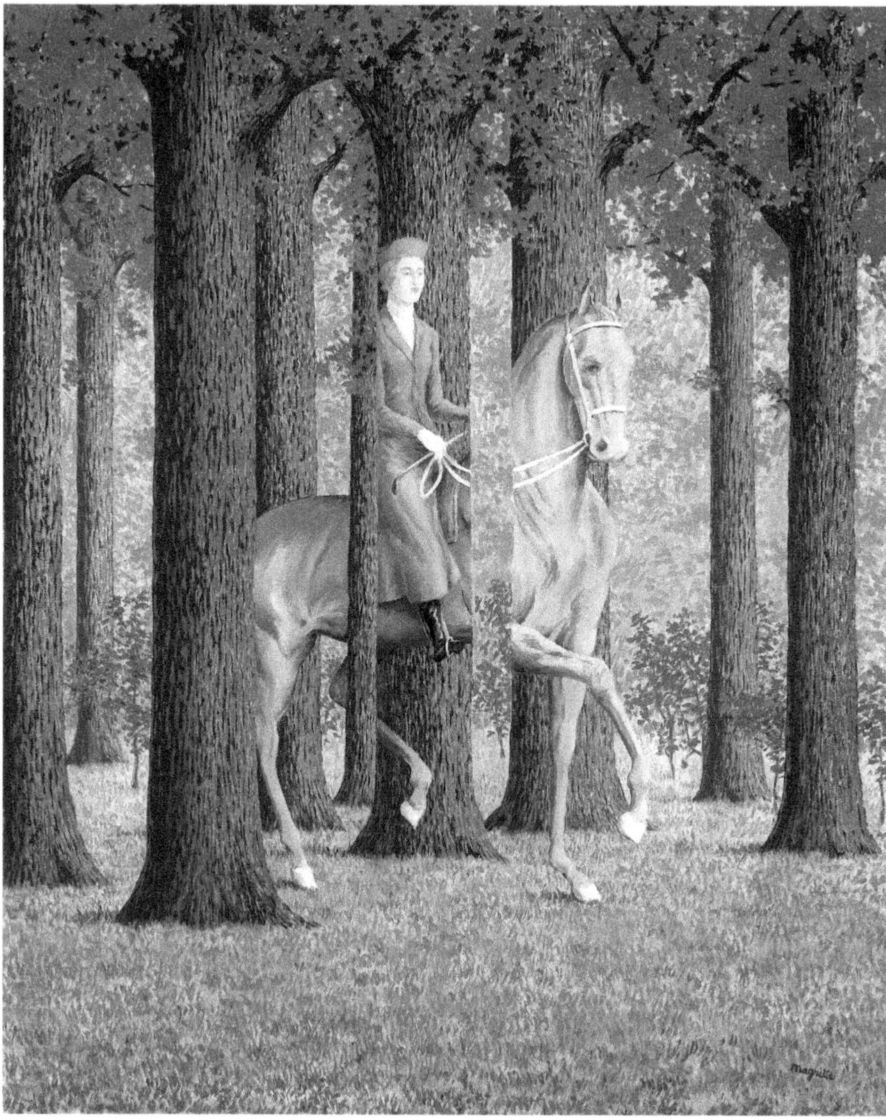

7.2 René Magritte, *Le blanc-seing*, 1965, oil on canvas, 32 × 25 ⅝ in. (81.3 × 65.1 cm). Collection of Mr and Mrs Paul Mellon. Image courtesy of the National Gallery of Art, Washington, DC

In regard to this second culture of mimicry, too, we can find some impressive examples among the Surrealists—and the list here is much longer. Certainly, one has to mention Roland Penrose, who actually wrote a *Home Guard Manual of Camouflage* during the Second World War. But perhaps the closest follower of Thayer's principle among Surrealist painters was, again, René Magritte, who in his later painting *Le blanc-seing* (1965), fragmented a rider and her horse in a wood in a similar way as to how one would see a zebra in front of a black background (Figure 7.2).

In a picture like this, again, questions about likeness and representationalism do not count as an uncontested frame of reference. What counts is "dazzle"—as it does in Thayer's ruptive brand of mimicry. The point is how perception works when representation fails and negation becomes impossible. I consider this Thayer-influenced branch of Surrealist painting not only to be much more Darwinian, but also to be much more interesting than the first one—and with this in mind, I wish to conclude with a most prominent example of non-representationalism in a Surrealist work: Salvador Dalí's and Luis Buñuel's movie *Un Chien Andalou.*

A "coup d'œil"

There are only very few special effects of the 1920s that still have a visual impact on audiences accustomed to current aspects of filmmaking. One of them is certainly the razor-blade cut of the eye in the opening sequence of *Un Chien Andalou.* The striking feature of this special effect is that it seizes the attention of the spectator in such a way that they do not see the most simple fact that it is *not* the eye of the lady presented that is cut. It is, indeed, not even a human eye, but most visibly the eye of a cow. Nevertheless, the cut and the montage direct the perception in such a way that even this knowledge does not make a great difference. And this is what makes the cut a fact of filmic mimicry. The film—like natural mimicry—aims at emotional perception. It does not so much deal with representational kinds of disguise or dissimulation. It rather stages one of the strongest special effects in the history of cinema. And as such, its aim is a production of presence, instead of communication of meaning.[22] Perception is caught in a state without "as if" or negation. The option of saying, "This is not an eye, but a representation of an eye," does not even really help insulate oneself against the shock—it can only make it less manifest and more subliminal. And the same goes for saying, "This is not a human eye." This dimension of special effects leads to Coleridge's famous concept of the "willing suspension of disbelief" as a basis for fictional aesthetics *ad absurdum*. People who watch this clip generally share an incredible pain at what is presented. Most of them fail to *establish* disbelief. Or, to say it more precisely: they fail to establish an order of perception where belief (in the "reality" of a scene) counts as an uncontested frame of reference—just as in the moment of jumping away from a snake, the fact of realizing it is a snake does not play a role. The special effect takes place where ironic distance, "as if," and negation simply have no place.

Mimicry here is not *represented as* mimicry (as would be the case when showing a mantis—the very insect many Surrealists were so fond of). Nor does the film represent likenesses and hence use semiotic kinds of mimicry. It does not present similarities or disguises or masquerades to the detached eyes of an aesthetically engaged observer—as Nabokov wished to think of it. What Dalí and Buñuel do instead is to use mimicry in a similar way as to how it works in evolution. They make viewers perceive, feel, and act at the very same time. They link a scene quite viscerally with the closing of one's own eyes and with the embodied emotions of terror and disgust. They make perception share a situation that will only begin to present itself as a ruse or masquerade once one *is* detached and once one *does* analyze: But it is impossible to analyze the film and thereby "resolve" the very state of the immediate impact of the scene at the time of viewing. At the instant the scene plays, as mimicry works, there is no such thing as an "as if," and there is not even such a thing as representation. There is presence. Mimicry here, like in nature, does not work as a fact of disguise—it works as a special effect —and one could even go so far as to say that special effects in general rather function like mimicry in evolution than as facts of (representational) mimesis. And as we have seen, the same can be true for cultural products too. Only when we begin to understand the mimicry of special effects as a deception do we begin to think of it in terms of likeness.

This brings us to the blurred moment of a third type of illusion, besides the "symbolic" illusion and the "iconic" illusion: the indexical illusion. Unlike the blurred moment I experienced with the "words" "spoken" by the mynahs, things in this case are not about the threshold between an icon (birds mimicking humans) and a symbol (a phrase). And unlike Magritte's pipe, it is not about the threshold between an index (a pipe as an object of perception) and an icon (the representation of a pipe). It is about the coming into being of an index. This blur considers on the one hand the cut through the eye as an event shared as a perception in action and the cut of the eye as an "index" of pain, disgust and shock. This breaking down of a threshold sets mimicry on the verge of semiotic and non-semiotic perception. You cannot (as does a semiotic approach) just start with subject and object—even though you *can* describe mimicry in these terms and then talk about a disguise based on likeness. It is much more fruitful to start with enactive constellations and with a likeness beyond reference (which was also Breton's point). And the illusion involved here marks the verge of this very threshold between perception in action and perception for the production of meaning. We have to take into account not only the thresholds intrinsic to semiotics, but also the threshold towards any given semiotic order. We have to take perception into account where it has nothing to do with signification, not even the kind of signification where things simply stand for themselves. We have to take into account that referentiality is not simply a given fact, but something that even interferes with enaction. A useful account of mimicry, therefore, cannot limit itself to how signs work, but must rather consider how bodies work, too—even where signification is not the main issue. And it must account for the co-presence of seemingly distinct levels of signification. It must consider how symbols work as icons (the mynahs and me), how icons work as indexes (Margritte's pipe), how indexes work as enactions (mimicry of cutting an eye)—and finally, to close the circle: how enactions work as symbols.

The twist in this last option might seem a bit of blurring itself. But let me return to the cow's eye and deepen this semiotic account by entering into the domain of symbols as well. Stuart Liebman[23] has convincingly argued that Dalí and Buñuel staged a kind of simulation of a Freudian dream work, especially focusing on presenting a "latent" dream text in images. What he discovered was that most of the shock sequence of this movie was based on the puns of cuts, eyes, cows, and asses. The terms involved are "coup" (punch or cut), "kuh" (German for cow) "cul" (ass), and "œil de cul" (anus). Indeed, if you "parle français comme une *vache* espagnole" (speak French like a Spanish *cow*)—and the two artists *were* Spaniards—then you will pronounce "cul" just like "coup" and "kuh." The major pun at work in the very cut of the eye, though, is the "coup d'œil", the glance or glimpse—but literally the punch or cut of the eye. Once objectified, it can become a razor blade cutting through an eye.

Liebmann therefore concluded that the film works as a kind of rebus—i.e. a riddle, in which you always have to switch from icon-perception to symbol-perception, from looking at images for what they mimetically (re)present and for what they could mean if translated into language. This solution makes a riddle on the threshold of symbolic and iconic meaning out of the film—in the same way as Magritte's *Trahison des images* can be solved as a riddle concerning indexical and iconic meaning. But as in Magritte's case, the riddle-like mode of "reading" cannot account for its unsettling momentum. A dream, too, has no option for negation. Moreover, it will be semiotic and signify something only when interpreting it. And it is only then that it might seem to possess a kind of allegorical "truer" meaning of a latent text: only then might it appear as a *rebus*. While dreaming, there is nothing but the sequence of images and their *intrinsic* (emotional) meaning.

What is more important, though, is that the "coup d'œil," the mere glance or glimpse, is also the very condition under which mimicry works as a special effect. The film uses a kind of dirty trick—or, to say it in French, "coup de vache" (cut of a cow) by presenting a cow being cut. The stunning issue about mimicry as a special effect is, therefore, the Darwinian way in which it works *despite* conscious analysis of the lies and disguises, the ruse of this *coup de vache*. The film, even on the condition that it can be read as a sequence of symbols and like a rebus, does not cease to work with a phenomenology of Darwinian, rather than semiotic, mimicry. It does, yes, disguise *words* and make a masquerade out of them—but it does so in a way that the threshold focused on is not the semiotic threshold between words and their phonological imitation (as in my case of the mynahs). The words work only in latency—while the manifest film (like Freud's manifest "dream content") incorporates their meanings into a logic that is not semiotic.

The film makes language latently follow a paradigm of Darwinian, rather than semiotic, mimicry. It makes *use* of words as symbols, turning them into a state where wordplay makes no difference. Unlike in Freud's dream analysis, the latent meaning is not "truer" or "more important" than the manifest content. In fact, the opposite is the case. These meanings do not even really count in terms of what the film is about. You can forget about wordplay and the film will not lose much in the process. The manifest content is thereby not even really a "content" (i.e. something mimetically represented); instead, the presumed "mimetic representation of the cut through a human eye" should rather be considered as an enactive mimicry taking place between the screen and the living bodies of the beholders—in a similar manner an enactive mimicry evolves between hoverflies and birds. A similar transformation of the phenomenology of imitation—a switch between mimicry (or non-representational mimesis) and representational mimesis (or "semiotic" "mimicry")—might also be at stake if human language oscillates between the mouths of humans and the beaks of mynahs, at least as long as one does not focus too much on one's own semiotic questions, as I did throughout this chapter.

Notes

1. Manee Archawaranon, "Vocal Mimicry in Relation to Social Dominance in Hill Mynah *Gracula Religiosa*," *Journal of Biological Sciences* 5 (2005), 771–5.
2. Brian M. Foss, "Mimicry in Mynas (*Gracula Religiosa*): A Test of Mowrer's Theory," in *Psychological Modeling: Conflicting Theories*, ed. Albert Bandura (New Brunswick, Canada: Transaction Publishers, 1971), 63–9.
3. R.R. Tenaza, "Wild Mynahs Mimic Wild Primates," *Nature* 259 (1976), 561.
4. Thomas Nagel, "What Is It Like to Be a Bat?," *Philosophical Review* lxxxiii 4 (October 1974), 435–50.
5. Jerome Lewis, "As Well As Words: Congo Pygmy Hunting, Mimicry, and Play," in *The Cradle of Language*, eds Rudolf Botha and Chris Knight (Oxford: Oxford University Press, 2009), 236–56 (246).
6. Alva Noë, *Action in Perception* (Cambridge, UK: MIT Press), 2004.
7. James J. Gibson, "The Theory of Affordances," in *Perceiving, Acting, and Knowing*, eds Robert Shaw and John Bernsford (Hillsdale, NJ: Erlbaum Press, 1977), 67–82.
8. William M. Ramsey, *Representation Reconsidered* (Cambridge: Cambridge University Press, 2007).
9. Michel Foucault, *Ceci n'est pas une pipe* (Paris: Fata Morgana, 2005).

10. André Breton, *Le surréalisme et la peinture* (Paris: Gallimard, 1979), 11; and for a discussion see Françoise Py: "L'œil sauvage—André Breton critique d'art," in *Europe* 743 (1991), 135–43.

11. André Breton, *Nadja* (Paris: Gallimard, 1964), 14–15.

12. Ibid., 15.

13. Shaun Gallagher, *Brainstorming: Views and Interviews on the Mind* (Exeter: Imprint Academic Press, 2009), 61–2. This process has often been discussed as merely a fact of "instinctive" behavior, and from his example this seems to be true. But a very similar phenomenology is also present in the actions of trying to avoid car accidents, for example, and here, evidently, a cultural re-formation of instinct is present as well.

14. For a broader discussion of non-representational mimesis see Jan Söffner, "Non-Representational Mimesis (Grönemeyer with Plato)." Forthcoming in a special edition of *Etnofoor*, dedicated to imitation.

15. For example, Charles Darwin, "A Review of H.W. Bates' Paper on 'Mimetic Butterflies'," in *Charles Darwin, Collected Papers*, ed. Paul H. Barrett (Chicago: University of Chicago Press, 1977), 87–92.

16. Peter Forbes, *Dazzled and Deceived—Mimicry and Camouflage* (New Haven, CT: Yale University Press, 2009), 6–70.

17. David Kirsh, "Metacognition, Distributed Cognition and Visual Design," in *Cognition, Education, and Communication Technology*, eds P. Gärdenfors and P. Johansson (Mawah, NJ: Lawrence Erlbaum Associates, 2006), 147–80.

18. See for example, Vladimir Nabokov, *The Gift*, trans. Dimitri Nabokov (London: Penguin, 2001), 345–58.

19. Roger Caillois, *Le mythe et l'homme* (Paris: Gallimard, 1987), 107.

20. Gabriel Tarde, *Les lois de l'imitation* (Paris: Éditions du Seuil, 2001).

21. Caillois, *Le mythe et l'homme*, 103.

22. Hans Ulrich Gumbrecht, *Production of Presence—What Meaning Cannot Convey* (Stanford, CA: Stanford University Press, 2004), 139–41.

23. "*Un Chien Andalou*: The Talking Cure," in *Dada and Surrealist Film*, ed. Rudolf E. Kuenzli (London: MIT Press, 1997), 143–57.

8

Contemporary art and the aesthetics of natural selection

Ellen K. Levy

To what extent can one associate aesthetic preferences with particular ways of thinking and feeling? This is the question with which art historian Mark Roskill concluded his book, *The Interpretation of Pictures*, in 1989.[1] Although he asked this question in relation to an Expressionist or Neo-Expressionist language of shapes and colors, I believe that it could be asked of other aesthetic approaches as well. The contemporary artworks explored within this chapter evoke the subject of Darwinian evolution and prompt a number of questions. Can the aesthetic choices of artists result in art works that engage ideas and even divergent positions about natural selection and adaptation? A related question concerns whether the visual expressions of similar concepts tend to share stylistic traits. A further, and admittedly more speculative, aspect of aesthetics involves the manner in which scientists approach and express research on evolution. In this essay, in addition to exploring the aesthetic decisions of contemporary artists based on their ideas about selection and adaptation, I offer statements of several pertinent scientists about their own aesthetic orientations on these topics.

Debates in evolutionary theory

Almost from the outset, opposing camps developed in evolutionary theory around adaptationism even among scientists who were otherwise generally accepting of Darwin's theory. Of recent Darwinian arguments on adaptation, philosopher of science Ron Amundson comments, "Phylogenetically evolved adaptations qua adaptations are the primary explananda of natural selection, the central mechanism of neo-Darwinian theory."[2] In this view, natural selection involves the presence of heritable variation, which can be acted upon by environmental demands, resulting in "states of adaptedness" from which a selection is made. He explains further that, by contrast, developmental biologists posit that constraints on the developing form of the organism channel evolution and place natural limits on adaptation. The constraint versus adaptation debate is in essence about the priority given to constraints on form (by developmental biologists) or to the forces of natural selection themselves as explaining the existence of particular traits. Some proponents of adaptationism view natural selection as the exclusive force of change and minimize developmental constraints. However, there have been attempts to synthesize these views, and a field of evolutionary developmental biology has emerged as conceived by Rudolf Raff and others.[3] For example, biologists often stress timing events that are set off during an organism's development. These changes in timing, known as "heterochrony" are regulated by genes that, in turn, trigger other genes. Raff has attempted to position heterochrony more broadly so that the concept becomes applicable to the evolution of the full range of ontogenetic processes, from embryogenesis through adulthood.[4]

Adaptationism disputes have also entered into the theoretical areas of brain development and human psychology.[5] According to David J. Buller and Valerie Gray Hardcastle, "Evolutionary psychologists claim that the human mind is a network of distinct, yet interacting psychological adaptations."[6] Leda Cosmides and John Tooby, both evolutionary psychologists, postulate modules for a broad range of cognitive and behavioral functions, viewing the brain "as a collection of dedicated mini-computers whose operations are functionally integrated to produce behavior."[7] In general, evolutionary psychology follows a computational theory of function, which conceives of the human mind as an information processing system.[8] Claims that the mind has a massively modular structure and that the modules deal with specific domains have generated much debate. Although a proponent of brain modularity, cognitive scientist Jerry Fodor views evolutionary psychology as extreme in its ideas of "massive modularity."[9]

To what extent can these differing views of evolution—developmental biology and evolutionary psychology—be considered to have aesthetic counterparts? One would anticipate that any literal one-to-one correspondence between the artistic visualizations and scientific concepts of either evolutionary camp would necessarily be forced. The correspondences one might expect would have more to do with metaphor and resonance of overall meaning. Yet the correlations are, surprisingly, more precise than one might expect as will become clear.

Art and evolution

Two current artistic approaches to the subject of evolution among many possibilities include use of a cellular automaton (CA) and a genetic algorithm (GA).[10] A CA can generate complex patterns from simple rules and has been used by artists such as Richard Purdy, Paul Brown, and Leo Villareal. The ways in which a CA might be considered analogous to aspects of developmental biology will be explored. Another group of artists, including Karl Sims, Christa Sommerer, and Laurent Mignonneau, has made art through the implementation of the GA. The GA is an algorithm that mimics some of the process of natural evolution. The ways it can and cannot model natural selection will also be examined. A third artistic approach, artificial intelligence (AI), will also be considered in a project by artist Bill Tomlinson. He and his team have used AI methodologies to model social behavior, through manipulating virtual representations, data-structures, algorithms, interrupt handlers, and prioritizations that are implemented in a game.

Developmental biology

Developmental biologists share a belief that organisms are shaped by constraints, universal laws of forms, rules of pattern formation, and emergent properties of complex systems. In general, they tend to downplay the roles of genetic drift (a change to the gene pool of a small population that takes place by chance) and adaptation in shaping the evolutionary history of organisms. As explained by mathematician P.T. Saunders, "Even if random genetic mutations were the source of all heritable variation, this would not mean that the variations themselves were random. They may be triggered by the mutations, but their forms are determined largely by the properties of the developmental system of the organism, i.e. the process by which it develops from single cell to adult. If we want to understand evolution, we have to understand development."[11] Morphologist D'Arcy Thompson and mathematician Alan Turing were among those adhering to structural concepts of self-organization during the first half of the twentieth century. D'Arcy Thompson used mathematical methods to model biological growth in his 1917 publication, *On Growth and Form*. It subsequently became clear that some aspects of growth were controlled by chemical messengers, and in 1951, Turing constructed a model of reaction-diffusion equations to account for structural patterns in biology.[12] With the influence of cybernetics and the discovery of the genetic code, the exploration of biological systems in computational terms was encouraged, leading to the utilization of CAs.

Linguist Noam Chomsky (who is considered a structuralist in his approach to cognitive science) has explained the origin of the concept of physico-chemical constraints:

> It has been recognized since the origins of modern biology that such constraints enter not only into the growth of organisms but also into their evolution, with roots in the earlier tradition that Stuart Kauffman calls "rational morphology"...John Maynard Smith and associates trace the post-Darwinian

reformulation back to Thomas Huxley, who was struck by the fact that there appear to be "predetermined lines of modification" that lead natural selection to "produce varieties of a limited number and kind" for every species.[13]

Stephen Wolfram's book *A New Kind of Science* (2002) builds on this historical approach. It is in a line of scientific inquiry about evolution that stems from Thompson, who reinforced thinking about biological systems in terms of fluid mechanics and parameters of growth. Wolfram explains that the processes of growth and development within the individual make it either easier or harder for particular structures to arise.[14] He defines cellular automata as self-organizing groups of cells that develop in accordance with rules; the cells form patterns based on the instructions received. Their behavior changes depending on neighboring cells. Scientists use them to study complex systems in order to understand interrelationships between the individual elements and the whole. Wolfram downplays natural selection, explaining that through utilizing local, rule-based principles, cellular automata might conceivably yield results akin to evolution and mutation. He believes that the CA mechanism can explain certain recurrent patterns in nature such as that of mollusk shells.[15] Critics of his viewpoint maintain that there is no actual proof that nature really behaves this way. In addition, they point out that even if this were the case, it might be that natural selection favored a CA mechanism over others.[16]

Artistic visualizations based on cellular automata

Wolfram's approach has lent additional incentive to the production of art utilizing CA (cellular automata) and generative principles. Artists working with CA, including Purdy, Brown, and Villareal, explicitly reference their scientific sources and define CA's range of stylistic and aesthetic possibilities through their art. Purdy depicted the results of applying one of Wolfram's set of rules in a group of encaustic paintings. The visual presence of Purdy's *90* (2003) is in its fusion of the ancient principle of encaustic with the patterning generated by cellular automata.

However, CA need not always result in either geometric patterning or static forms. For example, Brown's computer artwork is also generated by CA, but the patterns are fluid in contrast to Purdy's painting. According to Brown, *SAND LINES* (1998) is one of a series of kinetic works he made that used "a cellular automaton to drive a non-repetitive, time / screen-based work. The CA provides input to a process that converts the binary cell data to tile families that are then animated."[17] To create his work, Brown developed a "tile-based" image generating system, involving rotations and mirror rotations of a single tile (planar unit). Focusing on rules helped him avoid defining his art as "self-expression" and enabled him to attain a distinctive

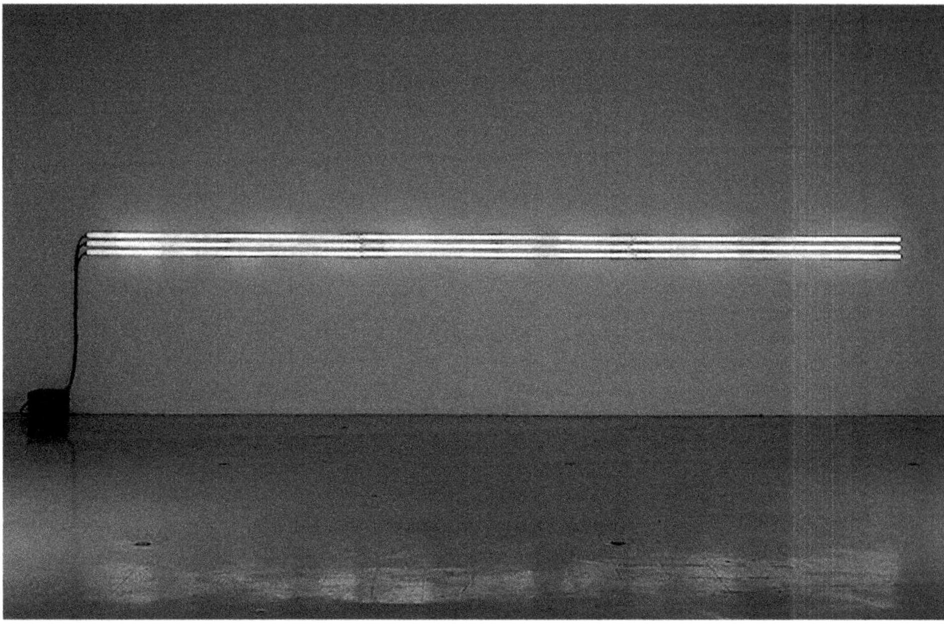

8.1 Leo Villareal, *Horizon*, 2004. LED tubes, custom software, electrical hardware, 12 × 288 × 4 in. Blanton Museum of Art, The University of Texas at Austin. Blanton Museum purchase with funds provided by the Brown Foundation. Image courtesy Gering & Lopez Gallery, New York. Photograph by Rick Hall

aesthetic that is characterized by the chosen method. As the artist stated, "In *SAND LINES* the image processing has been selected to consolidate the stone / sand identification and create a conflict with the animation which does things that the materials represented could not."[18]

Villareal uses CA rules to create light patterns in real time. *Horizon* (2004), is a 24-foot installation of glowing Plexiglas tubes filled with 3,888 red-green-blue light-emitting diodes (LEDs) that are individually modulated and produce an enormous variety of colors (Figure 8.1). Villareal was inspired by the "Game of Life," a cellular automaton program devised by mathematician John Conway in 1970.[19] Villareal uses software that utilizes a set of local rules to govern autonomous agents within a matrix. In Villareal's hands, this software results in a complex, kaleidoscopic visual experience.

Artists may use CAs for a variety of purposes (i.e. as a system to generate light patterns), but intentionally or unintentionally, they may also engage concepts of evolution because science has defined their use to a large extent. The CA offers a bottom-up way to create scientific models of biological phenomena and self-organization based on physical laws, albeit with limitations. CAs can utilize descriptions (computational rules) of the dynamic behavior of various cells and chemical reactions to model differentiation, mitosis, and mutation.[20] They can also emulate patterns on animal coats. A range of aesthetic effects can result regarding patterns, colors, and the clustering of forms. They can be displayed as still images or dynamically as part of sculptural installations.

Adaptationists

John Holland introduced GAs (genetic algorithms) in the US in the 1970s, and artists now use them widely. GAs are modeled loosely on the principles of evolution via natural selection. They work very differently from CAs and are used by artists much as industrial engineers have used them to design engines, but without the same practical aims. The engineers seek efficiency whereas the artists tend to explore the possibilities they permit. If an engineer wanted to augment the performance of a diesel engine, they would implement the GA in the following way: starting with a certain number of virtual engines that have different sets of traits, the traits are then mixed to produce another generation of engines, and survivors are selected to parent yet another generation. Some random changes (mutations) are added, and the process is repeated for many generations. The fitness criteria to enhance a diesel engine could include engine efficiency and minimal pollution. As a result, what might have taken generations to achieve under usual circumstances can, with the assistance of supercomputers, be done very quickly. The survivors of this virtual process are those that come closest to a desired result. The GA creates a population of solutions, and mutation and crossover operations can further the possibilities. In turn, these generate a diversity of solutions from which favored ones are selected, thus defining fitness.

To what extent can GAs be said to incorporate processes of "natural selection"? The designer writes a fitness function algorithm to determine whether members of a population will be favored through replication. The software offers total control of parameters including the fitness function, the form of the solution, the nature of the genetic operators, and the method of replication. As a result, this method makes minimal use of chance. Biologist Thomas Ray points out that the GA uses artificial selection rather than natural selection since the designer writes a fitness function algorithm that determines which members of a bit string population will be replicated.[21] Wolfram similarly points out that basic descriptions of natural selection do not correspond to a unique simple computer program. The difficulty with GAs revolves around eliminating less fit organisms. As another example, artist John Simon Jr. has designed artworks to try to answer the question whether natural selection can, in fact, be programmed into a computer, given that its essence is to allow for spontaneity, and be without purpose or design. Simon stated that he learned that a system designed by programmers rarely gets outside of itself, since, to create coherent images, they must supply a definition of the desired outcome to such a great degree that they are in effect just designing the images themselves.[22]

Visualizations based on genetic algorithms

Ray has collaborated with artists Christa Sommerer and Laurent Mignonneau to produce the artwork *A-Volve* (1994). The audience shapes the virtual creatures that are seen in a glass pool filled with water that is overlaid with a touch screen. The visitors design the virtual creatures with their fingers on the screen. The forms they design determine the capabilities of movement of the virtual creatures, which, in turn, defines their fitness; the fitter population has better mobility. In addition, an offspring can occur when two of the creatures meet through a "nature-like reproduction mechanism." Sommerer and Mignonneau consider the visitors as analogous to an external selection mechanism. They summarize that the creatures result through the "interaction of the visitors and the mating process of the creatures themselves. ...the system is able to evolve over time toward fitter populations."[23] These artists view their aesthetic as similar to that of John Cage, courting indeterminacy and chance. They have rejected the traditional idea of producing an art object in favor of a work constituted of the processes of nature and in which they, themselves, have given over their agency to an audience in interaction with a dynamic art form.

Ray similarly wanted to test a new range of aesthetic possibilities. He took a derivative of Karl Sims' *Evolved Virtual Creatures* (1994) and developed virtual organisms according to the aesthetic, emotional, and empathetic preferences of participants. Ray describes how, in Sims' *Evolved Virtual Creatures,* the computer program implements a virtual genetic system that generates a body built from blocks, along with a virtual neural circuitry to control movements of the joints. In essence, genetic algorithms generate the "neural systems" that control the "muscular movements" of virtual creatures that enable movements like swimming, walking, and jumping. A virtual genotype is comprised of a directed graph of nodes and connections that contains developmental instructions. The graph then specifies the phenotypes (the corresponding creature morphologies) that can result. The computational modeling of biological neurons allows for exploration of many kinds of activities. Ray's stated aim was to "evolve virtual pets to which humans can form strong emotional bonds. ...[believing] that this will lead to the evolution of creatures which evoke even stronger emotional, empathetic, and aesthetic responses."[24] Each "virtual pet's" genome contains a randomly determined number of boxes and defines a directed graph of connected boxes. The genotype is then translated into a phenotype by an algorithm that includes a limit on the total number of boxes generated. Ray's project gave viewers

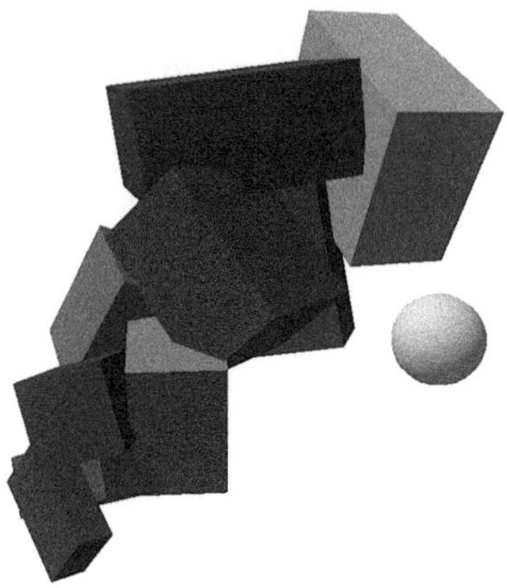

8.2 Thomas Ray, *AttractionDc13096*, 2000. Medium: capture of screen image. Photograph by Tom Ray

the ability to guide the course of evolution by allowing them to select "creatures" based on their emotional responses "in a fashion analogous to the breeding of animal pets." In Ray's words, "Rather than using GA selection for competition, I will evolve my creatures to love, by selecting for movement towards one another, then maximum surface contact, then continued relative movement while in contact."[25] Ray used a GA to evolve *AttractionDc13096* (2000). This "creature" appears to chase after an object that appears on the screen as a green ball (Figure 8.2). Like the anthropomorphic sculptures of Joel Shapiro, which are similarly blocky, rectangular forms are wedged together to create figurative works of a minimal aesthetic. They convey motion, energy, and attraction without being literally realistic and have the persona of cartoons or stick figures.

The aesthetics of cellular automata and genetic algorithms

The results of CAs typically evince dense, abstract patterning whether exhibited as the two-dimensional stills of a process that has occurred or as a computer work displaying a dynamic, ongoing process. By contrast, artists favoring interactive participation from viewers have been attracted to GAs. The Fluxus movement and performance art exist as precedents for this aesthetic approach, although one could argue that the elaborate coding required to produce such work is hardly improvisatory. Artists favoring GAs have often used naturalistic images such as plants as part of their installations as opposed to the geometric abstract patterning fostered by CAs. For example, Sommerer and Mignonneau selected real living plants to serve as an interface between the virtual artwork and the visitor in their work *Interactive Plant Growing* (1992). (Note that my generalization about naturalistic forms does not apply to the abstract volumetric forms developed by Ray.) Although somewhat over-simplified, it is as though one would be comparing a more formal visual aesthetic enabled by the artistic use of CAs to a post-modern sensibility that has tended to juxtapose real and virtual elements made possible with GAs. One could say that the comparison results in Modernism versus Post-Modernism, although there are clearly exceptions.

Theoretically, CA rules can be carried out without a computer. As artist Philip Galanter has pointed out, even systematic tiling systems constituted algorithms that existed long before there were computers. Generative systems provided artisans with a method similar to the use of CAs to create decorative murals.[26] Sol LeWitt's public artwork, *Whirls and Twirls* (2004), which was created for the MTA, continues a lineage of decorative tiling. The mural was executed in 250 porcelain tiles and six colors over a span that is 53 feet wide by 11 feet high.[27] To generate his prototype for the mural, assistants carried out the same designated set of rules. The MTA mural was fashioned by LeWitt's "aesthetic selection" from among a group of related drawings based on his assistants' interpretations of his initial rules. Broadly speaking, in art history style is generally regarded as both personal and collective, in that it can characterize an individual's approach as well as a creative ethos that helps define culture at a particular point of time. In the mural, cultural preferences can be detected on a broad level, as evidenced by LeWitt's choice of bright colors for the public space.

The benefits and limitations of analogy

Plant biologist Enrico Coen is among those forging links between adaptationists and developmental biologists (e.g. the field of evolutionary developmental biology). He compares the developmental processes of different organisms to discover how developmental processes evolved and to determine the ancestral relationship between them. Coen also examines how genes and growth interact to create specific shapes during development and how this is related to patterns of evolutionary diversity. In his book *The Art of Genes: How Organisms Make Themselves* (1999), Coen has memorably likened the processes of developmental biology to a range of artworks. He draws parallels between the interactive way an artist responds to a developing painting and the way "hidden colors" (properties of regions of cells or tissues) become apparent as "interpreting genes" respond to them, resulting in the formation of an organism's pattern. Coen relates the patterns in nature to human designs, pointing out that change comes about through a series of modifications, each reflecting what went on before. He explains, for example, that the bristle pattern of the fruit fly results from "scute" genes reacting to their ongoing, developmental environment. Coen analogizes the processes of evolutionary development to that of artist/tinkerers. But there are limits to analogies. Since scientists have largely defined the use of the CA and the GA for evolutionary studies, their use by artists often invokes concepts of evolution as well. Since the intentions of artists and scientists typically differ, it would be misleading to interpret the artworks as supporting a scientific position. However, the aesthetic differences and range of expressions permitted by the CA and the GA are very much to the point.

Scientific debates regarding brain development

The adaptationist controversy has resurfaced in speculations about cognitive development, including language. Chomsky has stated that scientists should consider applying factors of constraints and body plans even more widely to a range of developmental problems, including the function of cortical networks.[28] Chomsky and Stephen Jay Gould have further argued that language is a byproduct of other adaptations such as a large brain and not, itself, an adaptation.[29] Gould considered the brain to be subject to constraints at the levels of molecular architectures and size as affected by a growing organism. Changes in the developmental schedule were conceived as playing a central role in evolutionary development.[30] Biologist Annette Karmiloff-Smith has found evidence to support a "dynamic, epigenetic view of development" in which the timing of cortical events are important for the developing infant. The idea is that human brain structures occur in response to environmental inputs in real time with some predispositions.[31]

According to Jean Suplizio, what distinguishes the views of evolutionary psychologists like Cosmides, Tooby, and Steven Pinker from developmental biologists is their premise that the relevant internal mechanisms are adaptations.[32] They view human behaviors as the product of psychological mechanisms—"the interplay of many instincts competing"—that natural selection favored.[33] They direct our attention to the programs run by the brain and to its modularity, thereby leading to what their opponents consider to be a reductionist approach.[34] Evolutionary psychologists tend to view the mind as a set of information-processing machines designed via natural selection to solve adaptive problems faced by our hunter-gatherer ancestors.[35] Opponents of evolutionary psychology, including Gould and Richard C. Lewontin, have criticized the "adaptationist program" of E.O. Wilson and others. Many traits, they maintain, were actually byproducts, associated with adaptations, but not actually the result of natural selection.[36]

The aesthetics of neoteny

The last artistic example concerns the aesthetic visualization of the interaction of brain and behavior. Artist Bill Tomlinson has built a work that refers to the history and process of animal domestication, *AlphaWolf* (2002, Figure 8.3). His visualizations include child-like features on postures of the gray wolf (*Canis lupus*) apparently undergoing domestication. *AlphaWolf* was conceived as a way to explore social learning with several possible applications. A particular interest was how children would relate to it; and a stylized, expressive, and simplified cartoon image conceived as an artwork is presented to the user.[37]

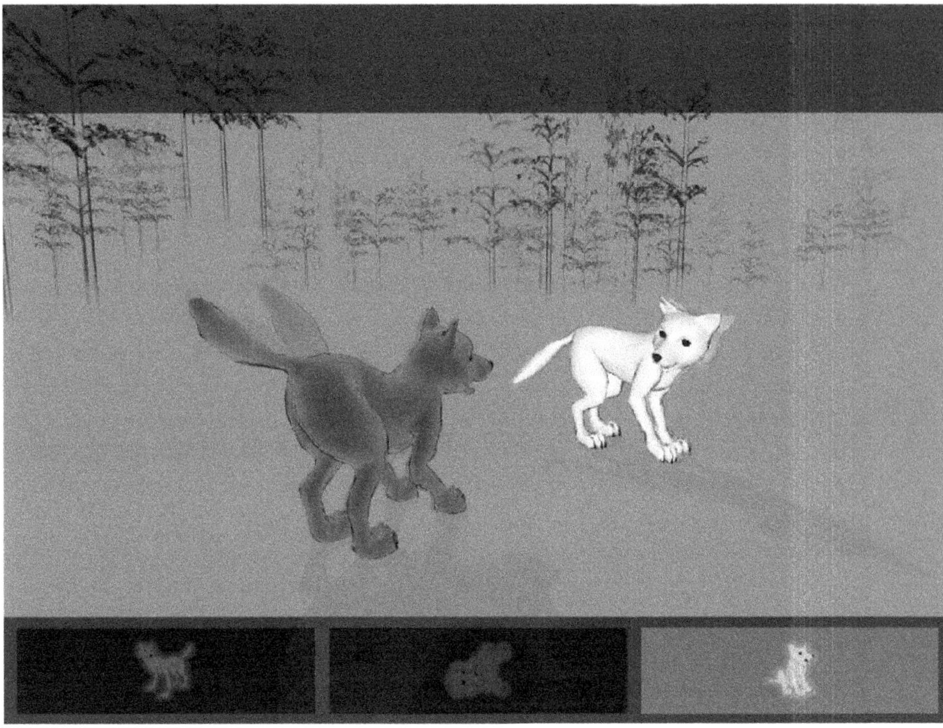

8.3 Tomlinson et al., *Two Virtual Wolf Pups Interacting in the AlphaWolf Installation* at SIGGRAPH, 2001. Medium: interactive digital installation. Dimension of original: variable. Generators of the image: Synthetic Characters Group, MIT Media Lab. Copyright 2001

Tomlinson and his team used AI to model animals that appear to be undergoing social integration. Each player is assigned the identity of a young wolf pup. Tomlinson solicits mimetic vocalizations of the ways wolves communicate from the viewers. The player "howls" at the wolves in the virtual pack. The aim is to help the assigned pup learn virtual behaviors that will establish its social order in the pack. By growling, whining, or barking into a microphone, each participant influences how their assigned pup interacts with the other members of the pack. The virtual pups, appealing to the viewer/player with their juvenile features, are programmed to "form" emotional memories among the pack that in turn have implications for future encounters.

The scientific process of domestication involves the following: a breeder selects for behaviors such as tameness and low aggression. The selection for these genetic traits in a wide range of domesticated animals may lead to morphological (phenotypic) changes (e.g. differences in fur coloration and increased skull size relative to body size).[38] Although the heritability of behavioral trait differences has been well established by now, the aim of some recent studies is to understand just how genes and molecules are influenced by developmental issues like timing, resulting in changes in entire suites of phenotypic traits. As one scientist noted, "Before our eyes, 'the Beast' has turned into 'Beauty,' as the aggressive behavior of our herd's wild progenitors entirely disappeared."[39] This observation tallies with Darwin's own experience. In looking at the work of animal breeders, Darwin realized that people could transform the appearance of animals under domestication. Observing artificial selection was a major impetus for his idea that nature could similarly mold creatures.

Some of the changes in bred animals appear to be related to neoteny (the retention of juvenile characteristics). Neoteny has aesthetic implications since many of the traits associated with youthful animals are considered aesthetically desirable. Gould has pointed our attention to a long list of juvenile traits like lack of coarse body hair and big heads that some find appealing.[40] The aesthetics of neoteny have long been explored in caricatures, political cartoons, comics, and more recently in *kawaii* (Japanese style of cuteness) and manga paintings.[41] Although he did not identify it as such, Thomas Ray examined the aesthetics of neoteny by deliberately setting out to generate a "loveable" (Ray's term) virtual blocky dog-like creature chasing a "ball."

Let me provide some additional background about domesticated animals since it touches on Tomlinson's project, involving vocalization as well as images. In the late 1950s, Dmitry Belyaev, a Russian geneticist, began a project to domesticate foxes. He later hypothesized (1969) that selecting and breeding wild animals for tameness would result in hormonal and neurochemical changes.[42] Recently scientists have carried out Belyaev's experiment and found that selective breeding among foxes (that did not include inbreeding) "resulted in 18 percent of foxes from the tenth generation demonstrating extremely tame behavior. Today elite (tame) foxes make up 70–80 percent of their experimentally selected population."[43] Researchers found that "selection for domestication effected changes in the timing of the postnatal development of certain physiological and hormonal mechanisms underlying the formation of social behavior." Subsequent experiments have suggested that vocal indicators for tameness and aggressiveness toward people are discrete phenotypic traits.[44] In other words, domestication was shown to select for behavioral traits. Belyaev's hypothesis proved to be accurate; not only were genetic changes apparent but neurochemical and hormonal changes occurred, such as drops in adrenal gland activity, which is associated with stress.[45]

The aesthetics of gaming

Tomlinson's virtual wolves have a repertoire of sounds and postures ranging from passive to aggressive. Darwin himself viewed expressive postures as

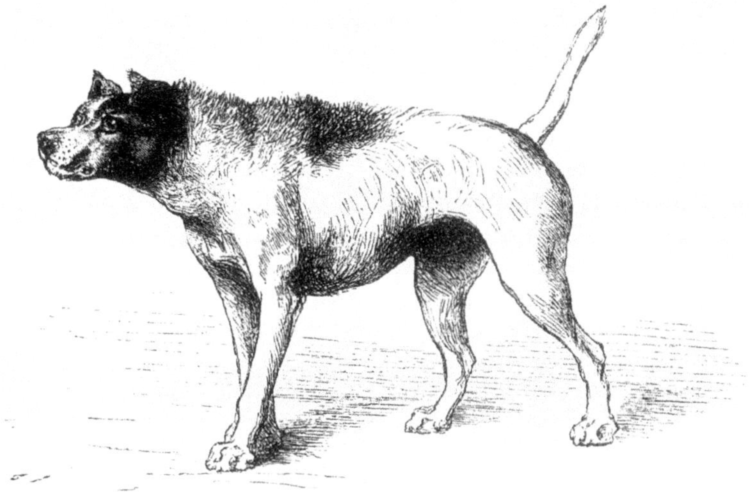

Fig. 5. Dog approaching another dog with hostile intentions.
By Mr Riviere

8.4a "Dog approaching another dog with hostile intentions. By Mr Riviere." Wood engraving. Figure 5 from Darwin, 1872, *The Expression of the Emotions in Man and Animals*, third edn (1998), with introduction by Paul Ekman (New York and Oxford: Oxford University Press), 57

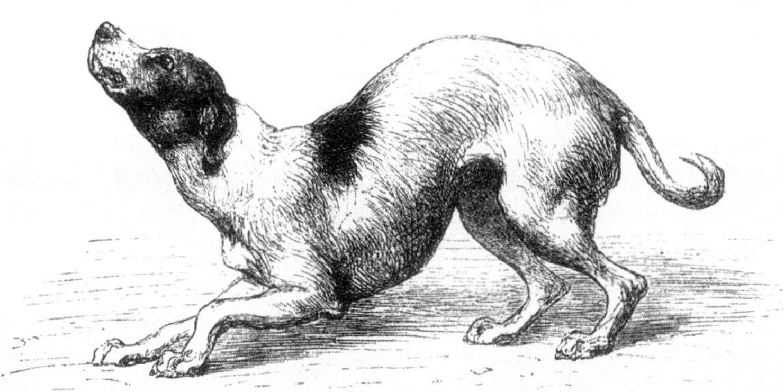

Fig. 6. The same in a humble and affectionate frame of mind.
By Mr Riviere

8.4b "The same in a humble and affectionate frame of mind. By Mr Riviere." Wood engraving. Figure 6 from Darwin, 1872, *The Expression of the Emotions in Man and Animals*, third edn (1998), with introduction by Paul Ekman (New York and Oxford: Oxford University Press), 57 instinctive physiological responses to an internal or external stimulus like threat (Figure 8.4).[46] Similar to Darwin's universalizing of animal postures, Tomlinson stresses the continuity between animal and human behavior through using body postures to visualize submissive and dominant behavior. The legend accompanying certain wood engravings for Darwin's *Expression of the Emotions in Man and Animals* identifies them as by "Mr Riviere [Breton Rivière]," a leading Victorian animal painter. In one case, the artist uses the ears and tail of the dog to signify its hostile intentions (top image); and in a related image (bottom), he portrays an affectionate frame of mind through the supplicating posture and curled tail.

Tomlinson's goal in *AlphaWolf* was to simulate the timing of innate behaviors (i.e. when they become active or inactive) in the context of a blend of hard-coded behaviors and learned behaviors.[47] This would seem to place his project on the side of the evolutionary biologists discussed earlier. However, Tomlinson also aimed to create virtual creatures with predispositions to behave in certain ways while retaining the ability to virtually "adapt" to their local environment. I believe that Tomlinson's work combines elements consistent with features of both structuralism and adaptationism. In this context it is interesting to note evolutionary biologist Juliet Clutton-Brock's observation that because the parent animals are forced to be reproductively isolated from the wild population, the actual biological process of domestication resembles natural selection.[48] Tomlinson's "artificial selection" relies on the vocalizations elicited from the participants of his game (often children).

Tomlinson and his team used psychologist Edward Thorndike's (1874–1949) "Law of Effect" as a model for his simulation.[49] The Law of Effect[50] was a neurological account of trial-and-error learning that Donald O. Hebb (1949) later reintroduced and expanded in a significant way, pointing to the brain's plasticity (the capacity to change with experience). Brain plasticity is essential for learning. Tomlinson used parameters in *AlphaWolf* (e.g. "ActionTuples" arranged into "ActionGroups") that determined which actions are mutually exclusive and which could be run simultaneously. In this way, his AI model attempted to capture temporal, situated, continuously changing information to account for the dynamic variability that reflected a branching web of possible solutions. Tomlinson's team effected a computational model of how social development and language might develop, applying insights from both cognitive science and behaviorism. Nevertheless, the model does not succeed in simulating our range of emotional experiences and would probably be of limited use in developing empathetic skills for autism (one of his aims). Because Tomlinson's goals were both aesthetic and scientific and his tools were those used by scientists, his aesthetic decisions meshed well with the scientific aims of analyzing developing behavioral skills in children.

Aesthetic biases of scientists

How can we link the research of some of the scientists we have noted to their own aesthetics? In his review of Wolfram's book in 2003, physicist Steven Weinberg stated:

> Wolfram is allying himself with one side in the ancient struggle between what (with much oversimplification) one might call cultures of the image and cultures of the word. In our own time it has surfaced in the competition between television and newspapers and between graphical user interfaces and command line interfaces in computer operating systems. …The ascendancy of the culture of the image has been abetted by computers and the study of complexity, which have made possible the simulation of complex visual images…I am an unreconstructed believer in the importance of the word, or its mathematical analogue, the equation.[51]

Weinberg suggested something important about the way different aesthetic orientations might lead scientists to favor certain approaches to problems in science. Weinberg himself analyzes nature through principles of physics that can be expressed mathematically and universally. In his own words, "It is the laws, not the phenomena, that interest us."[52] The criticism he levels at Wolfram is that reliance on computational modeling has caused Wolfram to conceive of nature as discrete, rather than continuous.

Wolfram has explained why visualizations are central to his theory, stating, "…while ordinary human language has little trouble describing repetitive and even nested patterns, it seems to be able to do very little with more complex patterns—which is in a sense why this book, for example, depends so heavily on visual presentation."[53] Wolfram describes his aesthetic as an elegance obtained from deriving complex visual patterns that may recall nature's own patterning from fundamentally simple rules. I suggest that this aesthetic approach is also prominent in the modernist patterns achieved by some of the artists using CAs.

Gould has also speculated about the role of aesthetics in scientific theory. He has claimed, for example, that Watson and Crick's discovery of the structure of DNA was due, in good part, to their hands-on tinkering with a cardboard model because there was a delay at the machine shop with the metal bases they had ordered.[54] In the article discussing this modeling, he concluded that, "Tinkertoys led Watson and Crick to the underlying shape of life's replication."[55] On another occasion, in his oft-cited "Spandrels of San Marco" argument, he framed his attack on the adaptationist program in terms of cultural complexity and the logic of architectural design. In the following passage he stated his own pro-visual biases claiming, for example: "[I]f texts are unified by a central logic of argument, then their pictorial illustrations are integral to the ensemble, not pretty little trifles included only for aesthetic or commercial value. Primates are visual animals, and (particularly in science) illustration has a language and set of conventions all its own."[56]

Thomas Ray addressed the aesthetics of evolution in a digital medium, pointing out that one would need to develop a new aesthetics to appreciate its products as being endowed with a beauty and richness of form and process.[57] His observation serves as a reminder that aesthetic taste is generally grounded in shared conventions; otherwise it would be impossible to even recognize originality.[58]

Conclusions

Many of the artists discussed are acutely aware of the issues posed by simulated evolutionary processes, frequently work with scientists, often sharing the same methodologies, and/or have had considerable scientific training. Their application of evolutionary schemas to their work cannot, however, be considered unbiased. Choices made from among many possible conventions, even when they involve mechanically applying a code, constitute aesthetic preferences and define personal expressions, although "crowdsourcing" used by some artists discussed above adds elements of improvisation to the final process. Chance, itself, thus finds an aesthetic basis.

Artworks embody a prioritization about what their creators have considered to be most important to attend to at a given point in time and history. An artwork that survives over a long period reflects its selection by a collective human aesthetic preference at particular points of time. The works in museums comprise a kind of collective "evidence" about these choices.

In response to the initial question, "To what extent can one associate aesthetic preferences with particular ways of thinking and feeling?" I conclude that these associations are not incompatible. In considering the question in the context of evolution, it became evident that each of the three approaches to simulation discussed (CAs, GAs, and AI) suggested a different conceptualization of natural selection. In addition, each kind of simulation enabled the expression of particular aesthetic qualities. The second question was whether similar styles would result from similar perspectives. The answer appears to be not necessarily, since artists working within each simulatory genre produced artworks that were very different from each other. The last question involved the manner in which scientists approach and express research on evolution. Scientists, like artists, have aesthetic preferences although they are not permitted the same range of expression. Style may nevertheless emerge in their approach to designing and carrying out a research project and formulating a hypothesis. For example, Tom Ray's "creatures" combined aesthetic curiosity with scientific research. As a final observation, in most processes involving "artificial selection" the main criteria that have found their ways into the specified fitness functions are typically aesthetic judgments. It appears that aesthetics trumps all.

Notes

1. Mark Roskill, *The Interpretation of Pictures* (Amherst, MA: University of Massachusetts Press, 1989), 102.
2. Ron Amundson, "Two Concepts of Constraint: Adaptationism and the Challenge from Developmental Biology," *Philosophy of Science* 61 4 (December 1994), 560.
3. Rudolph A. Raff, *The Shape Of Life: Genes, Development, and the Evolution of Animal Form* (Chicago: University of Chicago Press, 1996).
4. Rudolph A. Raff and Gregory A. Wray, "Heterochrony: Developmental Mechanisms and Evolutionary Results," *Journal of Evolutionary Biology* 2 6 (November 1989), 409–34.
5. David J. Buller and Valerie Hardcastle, "Evolutionary Psychology, Meet Developmental Neurobiology: Against Promiscuous Modularity," *Brain and Mind* 1 (December 2000), 307–25.
6. Ibid., 307.
7. Leda Cosmides and John Tooby, *Evolutionary Psychology Primer* 8. Accessed February 2011. http://www.infantcognitiongroup.com/LinkClick.aspx?fileticket=PVyY4NBKHGA%3D&tab id=118; also see Jean Suplizio, "On the Significance of William James to a Contemporary Doctrine of Evolutionary Psychology," *Human Studies* 30 4 (December 2007), 358.
8. Leda Cosmides and John Tooby, "Beyond Intuition and Instinct Blindness: Toward an Evolutionarily Rigorous Cognitive Science," *Cognition* 50, 1–3 (April–June 1994), 41–77.
9. Buller and Hardcastle, 310.
10. For the plurals of CA and GA, I use the designation CAs and GAs.
11. P.T. Saunders, "The Evolution of Form and Pattern, Art and the New Biology: Biological Forms and Patterns," *Leonardo* 22 1 (1989), 33–8.
12. See Andrew Hodges, *The Alan Turing Homepage*. Accessed December 30, 2010. http://www.turing.org.uk/turing/index.html.
13. Noam Chomsky, "Three Factors in Language Design," *Linguistic Inquiry* 36 1 (Winter 2005) 5.
14. Stephen Wolfram, *A New Kind of Science* (Champaign, IL: Wolfram Media, 2002), 387.
15. Ibid.
16. Steven Weinberg, "Is the Universe a Computer?," review of *A New Kind of Science* by Stephen Wolfram, *New York Review of Books*, October 24, 2002.
17. Brown, email correspondence with the author, July16, 2003.
18. Paul Brown, *Stepping Stones in the Mist*. Accessed December 30, 2010. http://www.paul-brown.com/WORDS/STEPPING.HTM; Paul Brown, email correspondence with the author, July 16, 2003.
19. Tracy Hummer, "Leo Villareal at Sandra Gering," *Art in America* June 2004; Martin Gardner, "Mathematical Genius: The Fantastic Combinations of John Conway's New Solitaire Game 'Life'," *Scientific American*, 223 (October 1970), 120–23.
20. G. Bard Ermentrout and Leah Edelstein-Keshet, "Cellular Automata Approaches to Biological Modeling," *Journal of Theoretical Biology* 160 1 (January 1993), 97–133.
21. T.S. Ray, "Evolution as Artist," in *Art @ Science*, eds Christa Sommerer and Laurent Mignonneau (New York/Vienna: Springer Verlag, 1998), 83.
22. John Simon Jr, email correspondence with the author, 2003; also see elaboration of these ideas in Ellen K. Levy, "Defining Life: Artists Challenge Conventional Classifications," in *Context Providers*, eds Victoria Vesna, Margot Lovejoy, and Chrisiane Paul (Bristol: Intellect Press, 2011), 279, 280.
23. C. Sommerer and L. Mignonneau, "Art as a Living System," in *Art @ Science*, eds Christa Sommerer and Laurent Mignonneau (New York: Springer Wien, 1998), 154.
24. Thomas S. Ray, "Aesthetically Evolved Virtual Pets," in 'Artificial Life 7 Workshop Proceedings'," eds Carlo C. Maley and Eilis Boudreau, *Proceedings of the Seventh Artificial Life Conference* (2000), 158–161. Accessed December 30, 2010. http://life.ou.edu/pubs/alife7a/index.html.
25. Ibid.

26. I thank Philip Galanter for this insight.
27. The designated location is the 59th Street–Columbus Circle subway complex in New York City.
28. Noam Chomsky, "Three Factors in Language Design," *Linguistic Inquiry* 36,1 (Winter 2005), 6.
29. Stephen Jay Gould, *Ontogeny and Phylogeny* (Cambridge, MA: Belknap Press, 1977).
30. S.J. Gould and R.C. Lewontin, "The Spandrels of San Marco and the Panglossian Paradigm: A Critique of the Adaptationist Programme," *Proceedings of the Royal Society of London B Biological Sciences* 205 11 (September 1979), 581–98.
31. Jean Suplizio, "On the Significance of William James to a Contemporary Doctrine of Evolutionary Psychology," *Human Studies* 30 4 (December 2007), 359; also see Annette Karmiloff-Smith, *Beyond Modularity: A Developmental Perspective on Cognitive Science* (Cambridge, MA: MIT Press, 1992).
32. Ibid., Suplizio.
33. Steven Pinker, *The Language Instinct* (New York: W. Morris and Company, 1994), 20.
34. Stephen M. Downes, "Evolutionary Psychology," *The Stanford Encyclopedia of Philosophy*. Accessed December 30, 2010. http://plato.stanford.edu/entries/evolutionary-psychology/.
35. Suplizio, "On the Significance of William James," 358.
36. Stephen Jay Gould, *Ontogeny and Phylogeny*.
37. Bill Tomlinson and Bruce Blumberg, "AlphaWolf: Social Learning, Emotion and Development in Autonomous Virtual Agents, Synthetic Characters Group," *AAAI Technical Report FS-01-02* (2001).
38. L.N. Trut, "Early Canid Domestication: The Farm-Fox Experiment," *American Scientist* 87,2 (March–April 1999), 167.
39. Ibid., 168.
40. Gould, *Ontogeny and Phylogeny*, 357.
41. D.H. Erwin, "The Goldilocks Hypothesis," *Science* 302 (December 2003), 1,682–3.
42. D.K. Belyaev, "Domestication of Animals," *Science Journal* (UK) 5, 47–52; D.K. Belyaev and L.N. Trut, "Some Genetic and Endocrine Effects of Selection for Domestication in Silver Foxes," *Journal of Genetics and Physiology* (1979), 416–26. D.K. Belyaev,"Domestication of Animals," *Science Journal* (UK) 5, 47–52.
43. Trut, "Early Canid Domestication," 163.
44. E.V. Volodin, A.V. Kharlamova, and L.N. Trut, "Vocalization Toward Conspecifics in Silver Foxes (*Vulpes vulpes*) Selected for Tame or Aggressive Behavior Toward Humans," *Behavioural Processes* 84 2 (June 2010), 547–54; also see S.S. Gogoleva, I. Volodin, A.E. Volodina, A.V. Kharlamova, and L.N. Trut, "Kind Granddaughters of Angry Grandmothers: The Effect of Domestication on Vocalization in Cross-bred Silver Foxes," *Behavioural Processes* 81,3 (July 2009), 369–75.
45. L.N. Trut, "Early Canid Domestication: The Farm-Fox Experiment," *American Scientist* 87,2 (March–April 1999), 166.
46. Charles Darwin, *The Expression of the Emotions in Man and Animals*, with introduction by Paul Ekman, rev. edn (NY and Oxford: Oxford University Press, 1998).
47. W. Tomlinson, B. Blumberg, and B. Rhodes, *How Is an Agent Like a Wolf?: Dominance and Submission in Multi-Agent Systems*. Accessed December 30, 2010. http://characters.media.mit.edu/Papers/MAMA-175.
48. J. Clutton-Brock, "Origin of the Dog: Domestication and Early History," in *The Domestic Dog: Its Evolution, Behaviour and Interactions with People*, ed. J. Serpell (Cambridge: Cambridge University Press, 1995).
49. Bill Tomlinson and Bruce Blumberg, "AlphaWolf: Social Learning, Emotion and Development in Autonomous Virtual Agents, Synthetic Characters Group," *AAAI Technical Report FS-01-02* (2001).
50. See http://www.education.com/reference/article/thorndike-edward-1874–1949. Accessed December 30, 2010.

51. Steven Weinberg, "Is the Universe a Computer?," review of *A New Kind of Science*, by Stephen Wolfram, *New York Review of Books* October 24, 2002. Accessed December 30, 2010. http://www.nybooks.com/articles/archives/2002/oct/24/is-the-universe-a-computer/?page=3.

52. Ibid.

53. Wolfram (2002), 631.

54. S.J. Gould, "The Shape of Life," *Art Journal* 55,1 (Spring 1996), 44–6.

55. Ibid., 46.

56. S.J. Gould, *Time's Arrow, Time's Cycle* (Cambridge, MA: Harvard University Press, 1987), 18.

57. T.S. Ray, "Evolution as Artist," in *Art @ Science*, 90; also see Karl Sims, "Evolving 3D Morphology and Behavior by Competition," in *Artificial Life IV Proceedings*, eds R. Brooks and P. Maes (Cambridge, MA: MIT Press, 1994), 28–39 and see Karl Sims, "Evolving Virtual Creatures, Computer Graphics," *SIGGRAPH* 94 Proceedings (New York: ACM, 1994), 15–22.

58. J.P. Crutchfield, "Is Anything Ever New? Considering Emergence," in, *Complexity: Metaphors, Models, and Reality, Santa Fe Institute Series in the Sciences of Complexity XIX*, eds G. Cowan, D. Pines, and D. Meltzer (Boulder, CO: Westview Press, 1999).

Index

References to illustrations are in **bold**

For Product Safety Concerns and Information please contact our EU representative GPSR@taylorandfrancis.com
Taylor & Francis Verlag GmbH, Kaufingerstraße 24, 80331 München, Germany

www.ingramcontent.com/pod-product-compliance
Ingram Content Group UK Ltd.
Pitfield, Milton Keynes, MK11 3LW, UK
UKHW022125190926
475702UK00013B/1647
* 9 7 8 1 1 3 8 2 5 2 6 7 7 *